Melville, Mapping and Globalization

Continuum Literary Studies Series

Related titles available in the series:

Coleridge and German Philosophy by Paul Hamilton
Fictions of Globalization by James Annesley
Mapping World Literature by Mads Rosendahl Thomsen
The Measureless Past of Joyce, Deleuze and Derrida by Ruben Borg

Melville, Mapping and Globalization

Literary Cartography in the American Baroque Writer

Robert T. Tally Jr.

continuum

Continuum International Publishing Group
The Tower Building
11 York Road
London SE1 7NX

80 Maiden Lane
Suite 704
New York NY 10038

www.continuumbooks.com

British Library Cataloguing-in-Publication Data
A catalogue record for this book is available from the British Library

ISBN: 978-0-8264-7151-2 (Hardback)

Library of Congress Cataloging-in-Publication Data
Tally, Robert T.
Melville, mapping and globalization: literary cartography in the American baroque writer / Robert T. Tally, Jr.
p. cm.
Includes bibliographical references and index.
ISBN 978-0-8264-7151-2
1. Melville, Herman, 1819–1891—Criticism and interpretation.
2. Setting (Literature) 3. Space and time in literature. I. Title.

PS2387.T35 2009
813′.3—dc22

2009012058

Typeset by Newgen Imaging Systems Pvt Ltd, Chennai, India
Printed and bound in Great Britain by the MPG Books Group

For Reiko

Contents

Acknowledgments

My thoughts on these subjects and the work on this book developed over many years, and I am grateful to all who have helped me in various ways during that time. I have benefited from the insights and inspiration of a number of professors, and I am especially grateful to Paul A. Bové and Jonathan Arac, who helped me to clarify and refine my initial ideas and who have continued to aid me over the years, offering advice on any number of subjects, as well as educating me by the example of their outstanding scholarship and criticism. They have been *mentors* in the finest and broadest sense of the word. Many other teachers deserve acknowledgment as well. Fredric Jameson helped to expand my intellectual horizons and offered careful, constructive criticism of my work, and his own writings have consistently provoked my thinking. Additionally, I received valuable encouragement along the way from John Beverly, Larry Goodwyn, Carol Kay, Ronald Judy, Toril Moi, Valentin Mudimbe, Dana Polan, Rick Roderick, Jim Rolleston, and Kenneth Surin.

Melville's life and work provide an apt, if not always pleasant, example of the conflicting pressures of economic necessity and intellectual ambition, of working life and the life of the mind. In my own, lesser voyages, I experienced a professional culture quite different from my current one, but one which helped me hone different skills and develop new ways of looking at the world that now serve me well in literary and cultural studies. Although working in a multinational, corporate law firm is not analogous to having one's hand in the tar bucket, I did gain a vista to regions previously unknown to me, enhancing my understanding not only of law and of finance, but precisely of "the world we live in," as C. L. R. James put it. Several people offered friendship and guidance during this epoch, among them Jyotirmoy Banerjee, Tor Brodtkorb, James Brown, Susie Bryant, Chris Edwards, Jie Jie Fertig, Sean Finley, Gary Carter, Josefina Melgar, and Holland West. In my continued wanderings, I found supportive colleagues in Nick Purrington and Ulda Datiles. Since my return to academe, Michael Hennessy has been an extraordinarily helpful departmental chairperson and friend. I have benefited from the wisdom of colleagues, including but not limited to Sue Beebe, John Blair, Lydia Blanchard, Kathryn Ledbetter, Priscilla Leder, Susan Morrison, Clifford Ronan, Teya Rosenberg, Jon Marc Smith, Susie Tilka, and Steve Wilson. Dean Ann Maris Ellis has also been a tireless supporter of my research. In 2007, I received a Research Enhancement Grant from Texas State University, which aided me in completing this book.

Before and throughout all of these experiences, my parents and grandparents have been there for me. Also, my life has been enriched by the cute perfection of Hillary L. Edmund, and more recently, by the antic *joie de vivre* of Dusty and Windy Britches. Above all, I owe the most to Reiko Graham, to whom this is dedicated and whose unflinching love and support made this possible.

Versions of Chapters 4 and 6 have appeared in slightly different forms, respectively, as "Anti-Ishmael: Novel Beginnings in *Moby-Dick*" in *LIT: Literature Interpretation Theory* 18.1 (January–March 2007): 1–19, and " 'Spaces that before were blank': Truth and Narrative Form in Melville's South Seas Cartography" in *Pacific Coast Philology* 42 (2007): 181–198. Some of the substance of Chapter 2 derives from my reviews of Jonathan Arac's *The Emergence of American Literary Narrative, 1820–1860*, published in *Amerikastudien/American Studies* 52.2 (2007): 249–251 and in *Nineteenth-Century Prose* 34.1/2 (Spring/Fall 2007): 410–419. I gratefully acknowledge the publishers.

Preface: "When Leviathan's the text"

An amusing scene early in *Moby-Dick* depicts the first meeting between Captain Peleg, one of the *Pequod*'s owners, and Ishmael, who wishes to join the whale-ship's crew. Peleg asks why the young man should wish to go a-whaling, and Ishmael responds, "I want to see the world." The old captain then directs Ishmael to the bow of the ship, to see what of the world he could: "Going forward and glancing over the weather bow, I perceived that the ship swinging to her anchor with the flood-tide, was now obliquely pointing toward the open ocean. The prospect was unlimited, but exceedingly monotonous and forbidding."[1] The joke, of course, is that open ocean is about all of the world that the crew of the *Pequod* are likely to see, but the image is also a fitting one in the early stages of the novel that is *Moby-Dick*. The novel itself is characterized by an unlimited prospect, but is sometimes exceedingly monotonous and forbidding. So is the world.

Moby-Dick, as Edward Said has noted (even with some embarrassment at the phrase), is really about "the whole world."[2] Certainly nothing Melville wrote before or after *Moby-Dick* offered the tremendous scope of this novel, and Melville's literary ambition was never so great as when he was writing it. Melville clearly had little interest in producing a simple narrative of a whaling voyage, an adventure story, or even the great American novel. Rather, as he allows his breathless narrator to put it in "The Fossil Whale," he attempted to produce a work of absolutely global scope, with an

> outreaching comprehensiveness of sweep, as if to include the whole circle of the sciences, and all the generations of whales, and men, and mastodons, past, present, and to come, with all the revolving panoramas of empire on earth, and throughout the whole universe, not excluding its suburbs.[3]

Moby-Dick is a global novel, in more ways than one.

Melville's desire to include *everything*—not to exclude even the suburbs of the universe—reveals not only his ambition, but the kind of project he undertakes in the novel. The description of the world and the desire to be absolutely comprehensive in the effort involves Melville in a fundamentally cartographic project. In *Mardi*, Melville had "chartless voyaged,"[4] but in *Moby-Dick* Melville includes a deliberative and intensive meditation on "The Chart," in which the "spaces that before were blank" are carefully inscribed and re-inscribed.[5]

Although one might compare any sort of writing to a form of mapping, as Peter Turchi has suggested in his book *Maps of the Imagination: The Writer as Cartographer*,[6] Melville specifically engages in a literary cartographic project to map the world. In an era of dramatic societal, economic, and cultural transformations, Melville develops a *literary cartography*, as I call it, in order to describe a world system in its emergence, a system which has (in the words of Fredric Jameson) become our own "untranscendable horizon" in the era of globalization.[7]

This world system was, in Melville's time, only *emergent*. The *dominant* force in mid-nineteenth-century American civilization was a type of nationalism that was embodied in literature and culture in the form of national narrative, even (and perhaps especially) as the actual nation-state known as the United States was approaching its greatest crisis in the years leading up to the Civil War. The fact that the "nation" was not a culturally consolidated power in the minds of Americans and others actually strengthened the force of national narrative, which after all becomes a means of shaping and empowering the national image. I argue that Melville's literary cartography of the emergent, *postnational* world system functions as an implicit, and sometimes explicit, critique of the national narrative of his own era.

In the twentieth century, American Studies established another sort of national narrative. Drawing on major literary works, many of which (like Melville's) stood in opposition to the nationalist rhetoric of their times, American Studies established a mythic history and culture for the United States, a national narrative involving a representative American subject in the American Adam, an exceptional American scene in the Virgin Land, and a more-or-less divinely ordained national mission in an Errand into the Wilderness. Using this "image repertoire" as Donald Pease has called it, Americanists relocated national narrative in great literary works (in what Jonathan Arac has called "the nationalization of literary narrative") and thereby transformed the literature into ideological embodiments of a nationalist tradition.[8] The very term *American Renaissance*, coined for better or for worse by F. O. Matthiessen's brilliant 1941 study, suggests a mixture of great national moment and aesthetic triumph. It is not a term used by Melville, but Melville becomes one of the chief figures in the American Renaissance and, by extension, a nationalist literary project in general. New Americanists, in their critique of such "old" Americanist interpretations, have often reinforced this practice by revealing these works as ideological embodiments of a nation now understood to be, among other things, racist, sexist, imperialist, and hegemonic. However the "nation" is defined, as the avatar of freedom and equality or as a force of repression, both interpretive traditions established literary works like *Moby-Dick* to be national narratives.

But Melville's work does not belong to the American Renaissance or to the national literature implied by such a label. Melville's work is *baroque*, directly at odds with the ideals of a renaissance, bizarrely and extravagantly exceeding the bounds of national narrative and cultural forms. Melville's baroque cartography imagines different spaces, and draws attention to the competing spatial forms of the state, the extraterritorial zones, the city, and the world, while also exploring the inter-relations among them. In so doing, Melville engages with a powerful sense of *world literature*, of a literature belonging to the world and no longer attached to the national at all. This world literature of Melville I find especially well suited to the task of thinking literature and culture in our own postnational world, in which the nation-state no longer stands as the uncontested model for social organization and in which the global circulation of power requires new ways of imagining our relationship to the world. American literary studies and the great American novel must play rather different roles in the era of multinational capitalism than in the "American Century." A postnational world requires postnational approaches to world literature, and the transnational turn in American Studies in recent years attests to this. New spaces call for new maps. The genius of Melville was that, in 1851, he detected the emergence of these new spaces and developed a baroque literary cartography to map them.

This book traces the movement of Melville's literary cartography and argues for its importance in a renewed program of American and world literary studies in the era of globalization. I begin by looking at Melville's baroque art, as well as his engagement with the social spaces and narrative forms in his own time. I show how his work, *Moby-Dick* in particular, embodies a postnationality, striving to register a global space and to criticize the nationalist forces in literature and culture. I provide a reading of the novel that registers this postnational aspect and which highlights Melville's nomad thought (to use Gilles Deleuze's term). In exploring Melville's work, I trace his development of a literary cartography, from his early personal narratives set in the South Pacific to his attempt to chart the urban spaces that were unsettling the maps of America and the world in the mid-nineteenth century. Both the form and the content of Melville's writing changes as he tries to make better and more powerful maps. I conclude by arguing for Melville's importance in literary and cultural studies in the era of globalization.

Melville's literary cartography is remarkable, in both its ambition and its scope. Melville really did want to create a global map that could incorporate everything, the large and the small, the individual and the collective, the residual, dominant, and emergent. As he notes in "The Fossil Whale," "Since I have undertaken to manhandle this Leviathan, it behoves me to approve myself omnisciently exhaustive in the enterprise; not overlooking the minutest seminal germs of his blood, and spinning him out to the uttermost coils of his bowels."[9] Melville's outrageously extravagant project, his baroque art and global literary cartography, disclose an almost hopeless desire for comprehension; the representation Melville attempts is quite impossible, but there is magnificent power in the attempt itself. Few writers in the nineteenth-century American literature essayed such a task, and no writers did so with the breathless intensity of Melville. As Melville insisted, the writing must expand to fit the subject, and the map must be designed to fit the space. If Melville had endeavored to produce a great national novel, the foundational text of a distinctly American literature, he could have organized his literary cartography accordingly. "When Leviathan's the text, the case is altered." Melville's manhandling of the Leviathan is a sign of our own grappling with totality, "its unshored, harborless immensities," in an effort to make sense of, and give form to, a world system that has exceeded the terms used to describe it.[10] In the baroque art of Melville's literary cartography, we may see a glimmer of our own impossible, but necessary, mapping project.

Chapter 1

Out of Bounds: Melville's American Baroque

Moby-Dick is, above all, a *baroque* text. It never belonged to any renaissance, American or otherwise, and all of its forces are mobilized in opposition to the national narrative implied by the discourse of cultural renascence. In its hyper-canonization, *Moby-Dick* has been accorded a special place in the American national literary culture, and along with *The Scarlet Letter*, it stands as the central novelistic text in the mid-nineteenth-century phenomenon dubbed "The American Renaissance." The label, of course, comes from the title of F. O. Matthiessen's field-establishing 1941 study, subtitled "Art and Expression in the Age of Emerson and Whitman," but which presents Melville as the final word on that age. In the twentieth century, Melville's work has been thoroughly canonized, in part through its supposed representativeness in an American Renaissance and in an American national culture. *Moby-Dick* is, as much as any, "the great American novel," a sacred text in the literary tradition that was formed partly in reference to it.[1] The idolatry of *Moby-Dick* has served both to establish American literary studies and to impress upon them a profoundly nationalist character. The great irony is that *Moby-Dick* so heartily resists the tradition into which it has been placed and in which it holds an exalted position. Both within the landscape of American letters in the mid-nineteenth century and in the more pervasively nationalist context of its status as literary idol in twentieth-century American Studies, *Moby-Dick* rejects its own canonization, and presents a thorough critique of the very processes it is thought to represent. As Gilles Deleuze has said about Spinoza's misplacement within a tradition of Cartesian thought by historians of philosophy, "there is no living corpse who raises the lid of his coffin so powerfully, crying so loudly, 'I am not one of yours.'"[2] So too does Melville's ghost cry out in opposition to the tradition of the American Renaissance. Melville offers something else, something vaster and darker, less comforting, and more grotesque: an American Baroque.

The concept of the baroque makes possible a new way of understanding *Moby-Dick* and Melville's other works, in the contexts of Melville's time and of our own, now in a postnational condition in twenty-first century.[3] By prying the novel loose from the tradition to which it has been consigned, I mean to produce a new reading that explores the degrees to which *Moby-Dick* struggles against the national narrative upon which American literary studies have been based. By virtue of its pre-eminent role in that tradition, its hypercanonicity and its status as a sacred text, *Moby-Dick* is uniquely well suited to undermine and revise that tradition.[4] *Moby-Dick* has long been viewed as a representative text of its own era, the era of national expansion, slavery, and civil unrest; moreover, in American Studies it has become the representative text of the nation. *Moby-Dick* has become a sacred text of the national literature, with respect to both the *national* and the *literary* aspects of the field: that is, as a distinctly *American* work that is also a distinctive work of *art*. There is certainly much in Melville's oeuvre to support a view of Melville as a literary nationalist, from Melville's ironic arguments that American readers must support American writers in "Hawthorne and His Mosses" to his panegyric to the American triumphalist spirits in "The Advocate" chapter of *Moby-Dick* and so on. But, in order to sustain this reading, one must also skirt around Melville's mocking, often vicious critique of the same nationalist imagery. The literary nationalists who successfully canonized and celebrated Melville had to shape from Melville's raw materials a particularly national literary figure. Such a tradition must ignore, suppress, and argue away those elements of Melville's work that mar that image. These *postnational* elements, as they might be called, form another kind of *Moby-Dick*. I am tempted to call it a "subterranean *Moby-Dick*" (after Deleuze's notion of a subterranean *Ethics* in visible in Spinoza's scholia),[5] but for the fact that it is not underground, but visible as daylight. Melville's work *is* postnational, and this aspect of the work is especially noticeable through its baroque extravagance.

The thesis of this book is that Melville, in his critique of the national scene of his own time, projected an image of a postnational world system, a world system that has become our own today. In my interpretation of *Moby-Dick* (among other works), I employ the figure of the baroque to exemplify its difference from the national literary tradition into which it has been placed and within which it has been valorized if not deified. Melville was out of place in his own time, the era of the American Renaissance, and Melville is an odd figure in the canon of twentieth-century American Studies. In many respects, Melville's nomad thought is perhaps better suited for our postnational world system today, in the twenty-first century, than for the period of national consolidation during his own lifetime or for the triumphal "American century" announced by and through American Studies in the twentieth century. In 1953, C. L. R. James declared that "The miracle of Herman Melville is this: that a hundred years ago in two novels, *Moby-Dick* and *Pierre*, and two or three stories, he painted a picture of the world in which we live, which is to this day unsurpassed."[6] On the heels of the Second World War and in the wake of the absolutely incontrovertible evidence of the destructive power of nationalism, including the nationalism at work in his own imprisonment at Ellis Island, where he wrote *Mariners, Renegades, and Castaways* while awaiting deportation as a subversive, James argues for a *postnational* Melville *avant la lettre*. James emphasizes his opposition to the national narratives of his own time even before beginning his argument, by including this as his original dedication: "For my son, Nob, who will be 21 years old in 1970, by which time I hope he and his generation will have left behind them forever all the problems of nationality." Clearly such problems have not gone away, but a world system in which the nation-state no longer functions as the central model for organizing power has emerged. With the advent of globalization, new problems arise. Just as James found in Melville the picture of the twentieth-century world, I argue that Melville's work produces a literary cartography of our world system, a late twentieth- and now early twenty-first-century era of globalization.

It is not that Melville is a prophet, however; it is that he sees in 1850 the outlines of a world coming into being. He sees a world system in emergence that has now, in the era of globalization, become our own. Melville's focus on the emergent globalization, as opposed to the then dominant nationalism or the increasingly residual forms of regionalism or parochialism, enables him to sketch a cartography of the world system.[7] Melville attempted to map the world system in its emergence, to identify the new spaces of conflict and liberty in an increasingly complex and unrepresentable world. In *Moby-Dick* especially, Melville's cartography is an attempt, in Fredric Jameson's words, "to think a system so vast that it cannot be encompassed by the natural and historically developed categories of perception with which human beings normally orient themselves."[8] To a certain extent, this has required new representational techniques: new ways of writing or of mapping the terrain. The term *mapping* is used figuratively, but only just so; the fact is that the attempt to represent the world system in its emergence is analogous to the attempt to map disparate places. Melville's baroque art is employed in a *literary cartography* of the world system.

This study of Melville's literary cartography necessarily implies a properly dialectical criticism, with negative and positive arguments. On the one hand, it contains a polemic, in which I argue that the Melville put forth and embraced by American Studies, and even by the New Americanists who have done so much to revise the assumptions and practices of the field in recent years, is not the proper Melville. This polemic involves a rethinking and rereading of Melville, specifically with respect to his dubious yet canonically established relationship to any nationalist project. I argue for a postnational Melville, one who sees and postulates the power of a multi- or supranational world system in which we live today. On the other hand, this study also presents a positive theory of the novel or of narrative as a form of mapping, of representing complex and global spaces in productive and telling ways. I argue that the baroque Melville is doing just that, and that the project of literary cartography is of central importance in literature, especially when considered in the light of what Jameson has called "that new spatiality implicit in the postmodern."[9] These two lines of thought interweave throughout this book and, I hope, come together to form a useful critical approach to literary studies in the era of globalization.

In this chapter I will discuss the concept of the baroque, examining the baroque character of Melville's work and exploring the ways that the age of the baroque establishes an image of modern society and literature, which in turn can offer glimpses of the emergent world system that has become our dominant frame of reference today. In Chapters 2 and 3, I look more closely at the relationship between narrative forms and social spaces, and I argue that Melville's literary cartography establishes a global space for world literature.

A Rough and Imperfect Pearl

In the United States, the cultural movement associated with both the dominant nationalism and emerging aesthetic of American literature in the mid-nineteenth century has been dubbed the "American Renaissance," the title of Matthiessen's monumental treatise. Although it was not his intent, Matthiessen demarcated the boundaries of a field of American literary studies and authorized a particular approach to the works studied, not to mention that the book also provided the title for a literature course taught at universities throughout the country. Matthiessen himself was not in favor of reading along nationalist lines or of isolating national literatures from one another, yet the effect of *American Renaissance* was just that: it reinforced the nationality and nationalism of the literature and it inspired subsequent generations to limit the geographical scopes of their researches to the beaches of the Atlantic and Pacific coasts. The book, and the conceptual emblem that its title soon became, dealt specifically with only five writers—Emerson, Thoreau, Hawthorne, Melville, and Whitman—and these only for a limited period (1845–1860, but focusing on the years 1850–1855 in particular); nevertheless, *American Renaissance* authorized the study of American literature more profoundly and more thoroughly than any other work before, and it established the framework for the field of American literary studies for a generation or more. Famously, Matthiessen did not himself accord the term *American Renaissance* the tremendous conceptual or categorical power that many of his followers would,[10] but even in expressing his doubts about the label Matthiessen indicates that these five writers, during these highly productive six years, produced a kind of renaissance in American culture akin to the flowering of artistic power seen in Europe during the period known as the Renaissance. As Matthiessen puts it, "It may not seem precisely accurate to refer to our mid-nineteenth century as a *re-birth*; but that was how the writers themselves judged it. Not as a re-birth of values that had existed previously in America, but as America's way of producing a renaissance, by coming to its first maturity and affirming its rightful heritage in the whole expanse of art and culture."[11] *Renaissance* thus retained its legitimacy.

Notwithstanding this odd disclaimer, Matthiessen was well aware that the word *renaissance* was a loaded term, likely to inspire rather specific feelings about the subject of his work. "Ever since the historiographic notion was elaborated by Michelet and Burckhardt—in 1845 and 1860, exactly bracketing Matthiessen's period—'renaissance' has carried with it a glamorous freight of secularism, progress, and pre-eminent individuality."[12] In other words, the notion of an American renaissance fit well into a larger national narrative, one developed during the nineteenth century and extrapolated in the twentieth by practitioners of American Studies. Perhaps against his own wishes, Matthiessen's label helped establish a profoundly nationalist enterprise. Applied to the rhetoric of an intensifying nationalism in the nineteenth-century United States, the term *renaissance*, with its particularly positive or even celebratory ring to it, becomes a vote of approval for a nationalist literary project. In the mid-twentieth century, the phrase carries an almost evangelical nuance, as the study of American literature comes to be associated with the proselytizing mission of transmitting American ideals and values to the rest of the world. As I have noted elsewhere, the nationalist literary project forms the basis for a well nigh religious belief in America, for good or for ill, in the discourse of American Studies established in the immediate post-World War Two years and still visible in the discourse of new Americanists critical of the discipline.[13] Notwithstanding the secularism implied by the term, who does not *believe* in a renaissance? The term is only used with approval.

The term *baroque*, which since its original coining has almost always carried a negative connotation, would appear to abrogate such a vote of approval. Originally a jewelers' term—the Portuguese *barroco*, referring to "a rough or imperfect pearl"—the term *baroque* was applied to the decorative arts at least as early as 1765, according to the *Oxford English Dictionary*. It was not used as a term of praise. As René Wellek notes, "In the late eighteenth and early nineteenth centuries the adjective 'baroque' was widely used as an equivalent of 'bizarre,' and the noun 'baroque' became established as a term for 'bad taste' in architecture."[14] By the late nineteenth century, German historiography had solidified the meaning of the term with its inherent sense of artistic decadence, specifically with respect to the period during which the unity of Renaissance art and architecture seemed to disintegrate. As Heinrich Wölfflin put it in his 1888 *Renaissance and Baroque*, a book which effectively standardized the use of the term, "It has become customary to use the term *baroque* to describe the style into which the Renaissance resolved itself or, as it is more commonly expressed, into which the Renaissance degenerated."[15] Although such usage would be expanded to encompass any period of aesthetic decadence, the *baroque* largely remained associated (and unfavorably contrasted) with the *renaissance*. Whereas the renaissance calls to mind formal ingenuity, rules, models, science and progress, the baroque is defined by excess, extravagance, anarchy, and ridiculousness. "Unlike the Renaissance, the baroque style is not accompanied by theoretical rules: it developed without models."[16] The relative formlessness of baroque productions, or better, the *inscrutability* of the baroque forms, underscored the negative impression of the baroque in general. Although the term no longer refers to "bad taste" *per se* by the late nineteenth century, *baroque* still carried an unfavorable connotation (if not denotation). "As an art-historical term, *baroque* has lost its suggestion of the ridiculous but its general use it still carries a suggestion of repugnance or abnormality."[17]

Similarly, in philosophy and literature, the baroque carries with it associations of excess and wildness, but also of degeneration and defeat. For example, Antonio Negri has suggested that the baroque itself must be understood as a period of reflection, meditation upon the failure of the Renaissance project to achieve its goals of a well laid out, rational, and harmonious system.

> Indeed, it is the very style of the epoch that changes. It is marked by the sense of defeat, by the state of separation in which the new man is forced to live, immersed in a nostalgia for the revolutionary plenitude of humanist and Renaissance development but caught up in the necessity of the crisis. The "baroque" seems to consist in the tension between the awareness of the severe, implacable defeat of the Renaissance on the one hand, and the always re-emerging nostalgia for that experience, on the other. Some have correctly defined the mentality that takes hold and spreads in this situation as "pathetic."[18]

In Negri's view, the baroque denotes a crisis in thought, one that inevitably leads to a reactionary political program intent on either restoring lost glories or at least retaining petty gains. However, the baroque nostalgia for the past also manifests itself in playful or parodic ways, as in *Don Quixote*, in which the "lost" realm of the chivalric romance is satirized as the modern world unfolds before it. Mikhail Bakhtin, for instance, has noted the "pathetic" quality of the baroque novel, yet he also argues that "Baroque pathos is determined by modes of *apologia*, and polemic. It is a prosaic pathos, one that continually senses the resistance offered by alien discourses, alien points of view; it is the kind of pathos associated with justification (and self-justification) and accusation."[19] The polemical spirit of the baroque suggests its power for a radical program of change, not merely a nostalgic feeling of ennui, but a revolutionary praxis. The disintegration or defeat of the renaissance project heralds the beginning of new social spaces, new relations, and new ways of representing them. In this sense, the *baroque* becomes a more positive label for transformative activities in art and politics.

The reappraisal of the term *baroque*, begun by Wölfflin and continued in the early twentieth century by Walter Benjamin and others, involved not so much a change in definition as a change in evaluation. Specifically, the characteristics of the baroque that had been cast in a negative light—ornateness, complexity, difficulty, extravagance—became valued as worthy artistic elements. For Benjamin, the baroque designated a period of decadence, yes, but also of abundant creativity, where artistic forces are unleashed and given free rein to produce hitherto unimagined works of art: "[T]he baroque is not so much an age of genuine artistic achievement as an age possessed of an unremitting artistic will. This is true of all periods of so-called decadence."[20] Such a view is worth bearing in mind when examining the "flawed" masterpiece that is *Moby-Dick*, which displays evidence on nearly every page of the author's "unremitting artistic will."

Moreover, as Benjamin's *Trauerspiel* study also attests to, the revalorization of allegory, following the long Romantic hegemony of the symbol, helped the baroque aesthetic return with new force in the twentieth century. With allegory, the baroque complexity of life is replicated while also being interpreted; the meaning of the baroque text is "expressed in a complicated configuration."[21] As Jameson has noted, commenting on Benjamin's theory, allegory is "the privileged mode of our own life in time, a clumsy deciphering of meaning from moment to moment, the painful attempt to restore continuity to heterogeneous, disconnected instants."[22] If this sounds somewhat like the modernist view of being and of time, that is not surprising. Indeed, although it might seem paradoxical and anachronistic, there is something rather *modern* about the baroque. Hence, it is not surprising that the term receives new and positive critical attention at the same moment that modernism in literature and the arts arises. John Beverley has suggested that the return of the baroque and its revalorization were contingent upon the crisis of liberalism around World War One and the Bolshevik Revolution. The baroque thus has a connection to modernism, at least insofar as modernists found in the arts of the baroque era a tradition on which to draw in creating their own responses to the historical crises of the day. "The emerging ideology of aesthetic Modernism [. . .] would recuperate the Baroque—and invert the Romantic hierarchy of symbol over allegory—in its own 'invention of tradition.' "[23] Just as the age of the baroque was marked by the disintegration of the Renaissance, the modernist moment saw the disintegration of the simpler world preceding it. As such, it makes sense that artists and critics struggling to shape their own responses to the current, unfamiliar conditions might look back to another period of decadence and creative explorations. In a world in which, as Marx and Engels had put it so movingly in *The Communist Manifesto*, "all that is solid melts into air,"[24] the degenerate forms of art might also yield the greatest opportunities for creative innovation and critique.

These characterizations of the baroque seem particularly apt when examining *Moby-Dick*. Despite its centrality to the project of the American Renaissance, *Moby-Dick* is really baroque: a rough and imperfect pearl, a bizarre or even ridiculous work of art, which has little respect for the formal limits of the traditional novel or romance. Indeed the baroque nature of *Moby-Dick* is visible in its extravagant form, in its willingness to stray from the story of one Ishmael and to include everything from everywhere, forming what was, in Evert Duyckinck's words, "a remarkable sea-dish—an intellectual chowder of romance, philosophy, natural history, fine writing, bad sayings."[25] The result is a complex, one might say *rhizomatic* work, extending in all directions and comprehending more that at first appeared possible. Art historian Henri Focillon has summed up the formal qualities of the baroque in a way that seems particularly suggestive of Melville's extravagance in *Moby-Dick.* As Focillon notes,

> In the life of forms, baroque is indeed but a moment, but it is certainly the freest and most emancipated one. Baroque forms have either abandoned or denatured that principle of intimate propriety, as essential aspect of which is a careful respect for the limits of the frame, especially in architecture. They live with passionate intensity a life that is entirely their own; they proliferate like some vegetable monstrosity. They break apart even as they grow; they tend to invade space in every direction, to perforate it, to become one with all its possibilities. This mastery of space is pure delight to them.[26]

Moby-Dick, too, delights in such mastery of space. Melville's general disrespect for the "limits of the frame" was noted with disapproval by more than one contemporary reviewer. For instance, one claimed that the novel was "distressingly marred by an *extravagant* treatment of the subject."[27] Literally *wandering out of bounds,* the "extravagance" of *Moby-Dick* is a primary characteristic of its baroque art. From its initial reception in 1851, to its rediscovery during the Melville revival, and to its eventual canonization as the Great American Novel—if not *the* Great American Novel, then at least one of the few top candidates—*Moby-Dick*'s form, or its proliferation of forms, has been the cause for both admiration and for concern. *Moby-Dick*'s extravagance is visible in its baroque excess, its disrespect for the rules of genre and the strictures of renaissance art. In adopting this baroque form, Melville signals his intent to master otherwise unmanageable spaces, and sets up his own literary cartography of the world system.

Crisis and Emergent Forms

As noted above, the term *baroque* is historically linked to the renaissance, in particular as a term of derision, denoting the degenerate cultural forms that arose after the glorious period which preceded it. With the transvaluation of the term in the early twentieth century, the term also achieved a sort of transhistorical meaning, uncoupling it from the Renaissance and opening it up to refer to any work that is extravagant, ornate, difficult and so on. But the particular period in European history (not merely an art-historical period) known as the "Baroque" is a crucial historical conjunction in which many elements of what would come to be understood as integral to modernity itself emerged.[28] The age of the baroque witnessed a breakdown of the Renaissance, not only in art and architecture, but in social organization and thought. Most significantly, perhaps, this moment marks the definitive waning of the Renaissance city-state and the rise of the modern nation-state form, with its ancillary and pervasive effects felt throughout Western societies and beyond. It also marks the advent of modern philosophy and literature, notably the birth of the modern novel. Itself a figure for the new and unfamiliar, the baroque is a sign designating the rupture between old and new, a space of becoming, or the critical moment of emergence.

Although one would not want to place exact geographical borders and firm dates around the baroque, the term is generally used to refer to sixteenth- and seventeenth-century Europe, and is often associated with the Counter Reformation, especially in Spain and Italy. But, rather than specifying a particular country or reign, the baroque is an international phenomenon, leaving visible traces on cultures throughout Europe and beyond. In *The Age of the Baroque,* Carl J. Friedrich limits his study to the years between 1610 and 1660, but it is clear than much before and after could be included. The baroque is a category well suited to the *longue durée* approach to history now associated with the name Ferdinand Braudel, since the most visible characteristics of the age cannot be clearly discerned for some time. As a historical label used to identify a particular way of thinking or set of events, it is not always very useful, but the baroque may serve as a convenient epithet for naming a number of related events and processes which reveal themselves in Europe at roughly this time. As Deleuze has put it, "The Baroque is a long moment of crisis."[29] This crisis can be detected in multiple sites.

The crisis that so many historians and critics recognize in the baroque epoch is more than just an adjustment of political or philosophical priorities or a momentary anomaly in history. As Wallerstein points out, "The term *crisis* ought not to be debased into a mere synonym for *cyclical shift.* It should be reserved for times of dramatic tension that are more than a conjuncture and that indicate a turning point in structures of *longue durée.*"[30] Among the elements of this critical turning point in world history, three fundamental spatiopolitical forms emerge during the baroque era, and these will define the contours of the modern world system. Most importantly, perhaps, the modern nation-state form emerges in the seventeenth century and becomes the dominant form for thinking and understanding the world system for the next 400 years. As part of its emergence, the extraterritorial zone of the periphery (most obviously the New World colonial possessions, but others as well) help to consolidate nation-state power while manifesting the state's centrifugal force. Similarly, the concentration of populations in cities, and especially in the great capitals, creates an urban space through the state's prodigious centripetal forces, establishing a model for the core, metropolitan power. These spaces, no less than the concomitant economic, political, and cultural forces that help bring them into being, mark the key geographic points of reference in a baroque world system and in modernity itself.

Friedrich argues that the key, defining aspect of the baroque age is the emergence of the modern nation-state. Although it is certainly the case that aspects of the state can be seen in earlier kingdoms or domains, the nation-state form emerges in the baroque. At the outset of the seventeenth century, as Friedrich points out, this emergence could not be taken as a foregone conclusion. For example, the leaders of the Counter Reformation certainly wanted to reunify Christianity in a strengthened Holy Roman Empire under the transnational leadership of the Pope. Yet, by century's end, the Pope's power over individual states (in terms of *Realpolitik*) was negligible, the Empire a mere adjunct of the Hapsburgs' territories, with independent and sovereign princes controlling Germany, and with Spain, France, England, and the Netherlands solidified as unquestionably modern states. The baroque moment of crisis gives birth to the quintessentially modern form, which will be the fundamental organizing force for populations, economies, social spaces, and even thought for the next four centuries, up to, in fact, our own time in the late twentieth- and early twenty-first centuries.

Though evocative, *birth* is probably not the best word to describe the emergence of the state form, since birth often implies a suddenness and a specificity that are not really applicable. There is no birthday for the modern state, although historians might agree on a given range during which it can be said to have come into being. *Emergence* is a better term, inasmuch as emergence also implies the coexistence of dominant and residual forms, often at odds with the emerging forms. Notwithstanding his own use of the term "birth" (*naissance*) or Nietzsche's (*die Geburt*), Michel Foucault explains that a genealogist is most interested not in origins *per se*, but in the emergence of phenomena or events amid "the hazardous play of dominations. Emergence is always produced through a particular stage of forces [. . .] Emergence is thus the entry of forces; it is their eruption, the leap from the wings to the center stage."[31] This leap to the center is precisely how the emergent forms of the baroque come to figure as the central elements of a modern world system. In language remarkably resonant with Melville's own baroque project, Friedrich explains why the metaphor of birth is inapt: "Such a great institution—and is not the modern state the greatest of them all?—'emerges' rather than being born, and what is meant by 'emergence' is the process by which during a given period its outlines become visible to all, like a whale coming to the surface of the sea."[32] The Leviathan of the baroque, Hobbes's in fact, emerges amid the crises of the age and transforms thought and society thereafter.

The emergence of the modern nation-state produces a particular social space and a particular way of thinking about it. As I will discuss in more detail in the next chapter, the rise of the state implies and makes possible the rise of national narrative and the ideological space that such narrative attempts to map. "The creation of strong states within a world-system was a historical prerequisite to the rise of nationalism both within strong states and in the periphery."[33] It is not that the nation is coextensive with the state, but a principal effect of the modern nation-state is to fuse the two older notions into a new, jointly understood phenomenon. The word *nation*, for instance, which had been used as early as the thirteenth century, generally referred to ethnicity rather than to any political affiliation or identity; after all, it derives from the Latin *natio*, or "birth." Overlapping usages were common, particularly in those European nation-states whose political identity was tied to a dominant ethnic group, and some versions of the older meaning still carry weight—today, for instance, in the nativist or sometimes downright racist views concerning who does or does not constitute a *true* American (or Englishman, or German, and so on)—even among its own citizenry. However, by the seventeenth century *nation* had begun to refer to the political formation of the modern state.[34] In Spain, there are the Spanish, never mind that they might also be Basque, or Jewish, or what have you. As Benedict Anderson has argued so persuasively, a nation is only, but also is powerfully, an "imagined community"; it is an imagined community closely associated with, if not entirely delimited by, a nation-state.[35]

Although not analogous to the state-formation of seventeenth-century European powers, the United States in Melville's time was undergoing its own crises of national and state formation. The crisis is visible not only in the most obvious areas of nineteenth-century American history, such as the tensions arising from westward expansion, the debates over abolition and states' rights, and the growing divide between industrial or mercantile interests and agrarian ones. Beyond these specific issues, there was no real national culture in the United States. The very term *United States* carried with it an awkward mixture of singular and plural, and it generally took the plural verb in antebellum writing, shifting to the ideologically significant though grammatically awkward singular around the time of the Civil War—i.e. "The United States *is* a great nation." In the United States, narrative was crucial to the process by which the American state developed its national identity. When national narrative first took shape in the 1820s and 1830s, "there was no fully operative national culture. National narrative was part of the process by which the nation was forming itself and not merely a reflection of an accomplished fact."[36] Although Henry Adams made the argument that the centralization of the American nation-state had occurred as early as the Jefferson and Madison administrations, by the Civil War and Reconstruction the nation-state's power was consolidated. In any event, the years during which Melville tried to work as a professional writer marked a central moment in the development of the idea of an American nation. As part of Evert Duyckinck's "Young America" circle,

Melville was himself involved in thinking what role literature was to have in the national system. Yet he was also among the most skeptical of national goals and rhetoric. The crisis of his age, like "the long moment of crisis" that was the baroque era, forced new thinking about the national, and about the forms associated with it in one way or another. The mapping of the newly recognizable yet still mutable national space would also require maps of other, related spaces as well.

As Wallerstein's analysis shows, the rise of the absolute state and mercantile capitalism caused, and are occasioned by, development in other geographical and economic areas. Wallerstein labels these areas *core, periphery,* and *semiperiphery*.

> World-economies are divided into core-states and peripheral areas. I do not say peripheral *states* because one characteristic of a peripheral area is that the indigenous state is weak, ranging from its nonexistence (that is, a colonial situation) to one with a low degree of autonomy (that is, a neo-colonial situation).
>
> There are also semiperipheral areas which are in between the core and the periphery on a series of dimensions, such as the complexity of economic activities, strength of the state machinery, cultural integrity, etc. Some of these areas had been core-areas of earlier versions of a given world-economy. Some had been peripheral areas that were later promoted, so to speak, as a result of the changing geopolitical of an expanding world-economy.[37]

Generally speaking, in the world system taking shape in the baroque era, the core refers to those states, chiefly Western European, that became central economic powers in the emerging world system of the sixteenth and seventeenth centuries, the periphery refers to those areas (such as Eastern Europe and especially Hispanic America) whose labor fed the wealth in the core-states, and the semiperiphery designated a middle ground, where aspects of both remained (as in parts of Italy or Germany not incorporated fully into one or the other zone). For instance, if the principal forms of labor in the core-states and in the periphery were free and forced or enslaved, respectively, in the semiperiphery an "in-between" form developed: sharecropping. These divisions roughly correspond to the spatiopolitical zones emerging in the early to mid-nineteenth-century United States.[38]

I am thinking of three zones in particular: the national, whose space is roughly coextensive with the imaginary geography of the state itself; the extraterritorial, which encompasses those far-flung domains (including, but not limited to, actual colonial possessions) that are imaginatively incorporated into the space of the nation while physically lying outside of it; and the local and especially urban zone, whose dense, concentrated space relies on the centripetal force of the nation-state but whose energies escape and exceed the nation's. These do not align directly with Wallerstein's categories, but mark the principal areas in the literary cartography or cognitive mapping of nineteenth-century American experience. In Chapter 2, I will discuss in more detail the mapping of these zones through the development of narrative forms in American literature, but before concluding this discussion of the baroque, it is worth looking at another critical form to emerge in the epoch: the modern novel, a form that also gives form to the world.

A Novel Form-Giving Form

Don Quixote stands at the threshold of modernity, tilting at the chimeras of a world he is entirely unprepared to face with a weapon of an era long past. In *Don Quixote*, the archetype for the modern novel as well as its first incarnation, an entire world emerges as another disappears. Don Quixote himself is the first homeless man, truly the embodiment of what Georg Lukács calls "transcendental homelessness," alienated from the world while ineluctably bound to it. His is a "world abandoned by God," in Lukács's elegiac phrase, and the principles that had organized that world can no longer obtain. In this baroque novel, the force of modernity is dramatized in ways that will also inspire Melville's own literary cartography. Don Quixote is lost in a new world system, in which the old categories are now devoid of their clarifying or explanatory powers. *Don Quixote* becomes a map, an ironic geography of the new spaces unfolding before our tragicomic hero's eyes. When Melville attempts to map his own world and the emergent spaces and forces of a critically different epoch, the Knight with the Sad Face's legacy is also visible.

The novel form is not necessarily limited to the modern novel, and *Don Quixote* has no definitive claim to being the first or most representative modern novel. Nevertheless, a long and venerable tradition has accorded Cervantes's masterpiece the honor, and the distinction between ancient and modern animates much of the traditional theory of the novel.[39] In his magisterial *Theory of the Novel*, Lukács contrasts the closed or integrated civilization of the epic world and the fragmented, modern world of the novel. The former hails from that "happy age" when "the starry sky is the map of all possible paths [. . .] The world is wide and yet it is like a home." This sense of being at home, of a harmony between one's innermost self and the cosmos, obviates the need for philosophy, according to Lukács. "For what is the task of philosophy if not to draw that archetypal map?"[40] On the other hand, the modern condition, whose representative form is the novel, is marked by a split between interior and exterior, a thoroughgoing sense of disorientation; it is "a world abandoned by God," and is characterized by a "transcendental homelessness," a feeling of *angst* that Heidegger would associate with the *unheimlich*, usually translated as "uncanny" but which of course also means "un-home-ly." As Heidegger makes clear in *Being and Time*: "In anxiety one feels '*uncanny*' [*unheimlich*] [. . .] but here uncanniness also means 'not-being-at-home' [*das Nicht-zuhause-sein*]."[41] This uncanny homelessness implies a spiritual and a physical sense of being lost (or "not *at home* in the world"). The age of the novel thus requires a map, a figurative way to connect one's self to one's world. And that is precisely what the novel becomes, a cartographic practice in which the writer produces a figurative or allegorical image of the world and one's place in it. The novel is a form that gives form to the world. It is a form of literary cartography.

It is not necessary that this literary cartography be limited to the modern world, although—as Lukács and others insist—the conditions arising from modern developments might make a more expressly cartographic project more desirable or even necessary. "The epic gives form to a totality of life that is rounded from within; the novel seeks, by giving form, to uncover and construct a concealed totality of life."[42] That is, the integrated civilization of the epic could be mapped as well, but it could be viewed as a relatively complete map, a world-picture that made sense, in part because it conformed neatly to lived experience. Modernity, at least as understood by this tradition in the theory of the novel, occasions a breakdown in the sense-making systems that had come before. The long moment of crisis that is the baroque is bewilderingly transformative, changing the geography of the world as well as the ways in which one understands the world. At the time Don Quixote christens himself and his steed and sets forth to battle the malevolent enchanters, giants, and rogues, a greater beast—Hobbes's Leviathan—emerges. Accompanying the rise of the state form are several other formations: an extensive geopolitical reorganization brought on by an accelerated exploration and colonization of the New World, the concentration of populations in the great capital cities, the development of an ever-more-refined capitalism and its attendant ideology of mercantilism, the skepticism and scientific inquiry that establishes "modern philosophy," and the development of a mass culture, facilitated by printing technology. This latter would prove comically disastrous to Quixote himself, who—at the beginning of Part II—must read about his own disappointments as he discovers that his story has become a bestseller. Small wonder then at the anxiety and sense of being "at sea" in such a new world.

The philosophical rupture that these new forms of social organization disclosed reveals itself in *Don Quixote* as well. Foucault asserts that "*Don Quixote* is the first modern work of literature,"[43] specifically contrasting this baroque text to the Renaissance. In the Renaissance world not yet abandoned by God, resemblance or similitude had constituted a semantic or logical coherence: the Almighty's signatures were still legible, so the world was a simple text to be read. But such an outmoded system of knowledge is the symptom of a diseased mind in Cervantes's novel. The madness of Don Quixote de la Mancha lies in his belief that resemblance or similitude may still hold the key to a knowledge of the world and a transcendent reality assuring its wholeness or integrity. Thus, a windmill might very well be a monstrous giant, a barber's wash-basin may in fact be the legendary Mambrino's Helmet, and droves of sheep may quite possibly be contesting armies. But by the seventeenth century, readers certainly knew better, and such a "reading" of the world appeared laughable. As Foucault explains,

> *Don Quixote* is the negative of the Renaissance world; writing had ceased to be the prose of the world; resemblances and signs had dissolved their former alliance; similitudes have become deceptive and verge upon the visionary or madness; things still remain stubbornly within their ironic identity: they are no longer anything but what they are; words wander off on their own, without content, without resemblance to fill their emptiness; they are no longer the marks of things; they lies sleeping between the pages of books and covered with dust. Magic, which permitted the decipherment of the world by revealing the secret resemblances beneath its signs, is no longer of any use except as an explanation, in terms of madness, of why analogies are always proved false. The erudition that once read nature and books alike as parts of a single text has been relegated to the same category as its own chimeras: lodged in the yellowed pages of books, the signs of language no longer have any value apart from the slender fiction which they represent. The written word and things no longer resemble one another. And between them, Don Quixote wanders off on his own.[44]

The epistemic break between the Renaissance and the baroque, as Foucault imagines it, is indicative of the spiritual condition of modernity which provides the ground for the novel form and the atmosphere in which the novel could thrive, as Lukács argues. Unlike the integrated or "closed" (*geschlossene*) civilizations that produce the epic, the disintegrated modern societies engender the novel form. "The novel is the epic of a world that has been abandoned by God."[45] The novel will have as its vocation the attempt to give some sort of reasonable form to a world no longer guaranteed of its recognizable contours by a transcendent reality. The chivalric romance, so brutally satirized in Don Quixote, "had succumbed to the fate of every epic that wants to maintain and perpetuate a form by merely formal means after the transcendental conditions for its existence have been condemned by the historico-philosophical dialectic."[46] That is, it becomes risible. The "transcendental homelessness" of the modern novel marks the burgeoning of a new era in which an entire system of meaning, hitherto maintaining itself in an integrated image of the world, was called into question.

Just as the novel itself is a form-giving form, a form that gives shape to an otherwise chaotic or disorganized ensemble of things, in the scientific realm, knowledge had to become the province of the knower: that is, it was insufficient simply to read God's script in His creation. Now the "reader," as in *Moby-Dick*'s "Doubloon" chapter, must "render" the text, interpret it via a more or less complex system of signs, create his or her own meanings, and attempt to make sense of or give form to the world. Hence the novel is an essentially epistemological form; like the map, it is a form of knowledge as well as an attempt to know. As Bakhtin points out, whereas in the epic (and in ancient literature more generally) "it is memory, and not knowledge, that serves as the source and power for the creative impulse [. . .] The novel, by contrast, is determined by experience, knowledge and practice (the future)." Bakhtin concludes by saying that, when "the novel became the dominant genre, epistemology became the dominant discipline."[47] In attempting to gain knowledge of the world while also describing it, the modern novel involves a cartographic project. The breakdown in legibility in Foucault's analysis of the baroque text might be likened to the spatial confusion or sense of being lost in a world that no longer makes inherent sense or lack suitable guideposts.

Jameson's "digression on cartography" in his discussion of postmodernism is instructive in this regard. Based on the pre-cartographic itineraries of individual travelers, ancient maps were often merely "diagrams organized around the still subject-centered or existential journey of the traveler, along which various significant key features are marked—oases, mountain ranges, rivers, monuments, and the like." The medieval discovery of the compass and the use of instruments like the sextant introduce a new dimension: "the relationship to the totality [. . .] At this point, cognitive mapping in a broader sense comes to require the coordination of existential data (the empirical position of the subject) with unlived, abstract conceptions of the geographic totality." Finally, with the development of the globe and the Mercator projection, "representational codes" intervene to disrupt the practices of "naively mimetic conceptions of mapping," as "it becomes clear that there can be no true maps (at the same time it becomes clear that there *can* be scientific progress, or better still a dialectical advance, in the various historical moments of mapmaking)."[48] Returning to "narrative" *per se*, then, one might also trace advances from the straightforward "truth-telling" testimony of individual eyewitnesses based on their necessarily limited empirical evidence to a broader conception of the individual's relationship to the unlived totality, and eventually to a conception of representational art itself, which must explore other fictional, perhaps fanciful, techniques in order to achieve its aims. One must project a world beyond that which one can experience, even if the projection required allegory or figuration. With Mercator's distorted map, the exaggerated representation of space—for instance, making Greenland as big as South America—serves the practical purpose of helping to navigate the actual space better, particularly with respect to determining longitude, the great terror of fifteenth- and sixteenth-century navigation. "True maps" of the old, mimetic sort, are less useful. Of course, all maps are fictions; even Borges's fabled imperial cartographers, who made a map coextensive with the space it purported to represent, were unable to create "true maps."[49] But with modern cartography, the fictions may also be necessary to the functionality.

Cognitive mapping, or the project of situating and representing oneself within a complex and seemingly unrepresentable social totality, is also an apt figure for narrative, which also tries to organize the data of experience and the larger totality into a meaningful ensemble. Jameson uses the term specifically in connection to the postmodern condition, globalization, and late capitalism, but it is clear that his model is not limited to these areas; indeed, Jameson has conceded that "cognitive mapping" is a modernist strategy and "code word" for a new kind of class consciousness. Jameson's model, although derived from Kevin Lynch's *Image of the City* and Althusser's revised notion of ideology as an imaginary solution to real problems, is largely rooted in a theory of narrative: narrative as a socially symbolic act, to use his own formulation in *The Political Unconscious*. (Indeed, Jameson's first use of the term comes from that book, prior to his essays on postmodernism, when he refers to realism as "a narrative discourse which unites the experience of daily life with a properly cognitive, mapping, or well-nigh 'scientific' perspective.")[50] Narrative is a fundamental way in which humans make sense of, or give form to, the world. In that sense, narrative operates much as maps do, to organize the data of life into recognizable patterns with it understood that the result is a fiction, a mere representation of space and place, whose function is to help the viewer or mapmaker, like the reader or writer, make sense of the world. Literary cartography, as I call it, connects spatial representation and storytelling. If narrative has always been a way of orienting ourselves, both in the relatively stable semantic universe of Lukács's closed civilizations or in the Quixotic modern world where the center does not always hold, then—in an era of globalization—the cartographic function of literature may be even more essential to the individual and collective reconquest of the *Lebenswelt*.

In Melville's writings, he sought to make sense of, and give form to, the world in which he lived, but he also, as C. L. R. James insisted, produced an image of the world we live in. I argue that he does so by exploring those emergent spaces and properties of a multinational world system, while also analyzing and offering critique of the dominant forms in his own day. He *projects* an image of the world system in its emergence. Projection, in literature as in cartography, allows one to give form to a world in order to make it legible. I think here of Oedipa Maas in Pynchon's *The Crying of Lot 49*, who, when facing a cripplingly complicated situation that seems to make no sense, says to herself, in language wholly appropriate to a discussion of Melville's own literary cartography, "*Shall I project a world*? If not project then at least flash some arrow on the dome to skitter among constellations and trace out your Dragon, Whale, Southern Cross. Anything might help."[51] Anything indeed. The constellation, composed of imaginary lines drawn between otherwise unrelated lights in the sky, is another kind of map, a baroque figure for a system almost too large to imagine, much less master. In Melville's work, the baroque excess signals an attempt to grapple with a system far exceeding the representational scope of certain narrative forms, and in manhandling the Leviathan, Melville endeavors to master, at least provisionally, the spaces of the emerging world system.

Chapter 2

Spaces of American Literature: Geography and Narrative Form

I take SPACE to be the central fact to man born in American, from Folsom cave to now. I spell it large because it comes large here. Large, and without mercy.

Charles Olson[1]

Upon first entering the study of the eponymous mansion in Hawthorne's *The House of the Seven Gables*, the reader is struck by two powerful figures, each framed upon the wall and each serving to frame the narrative as a whole.

> One was a map of the Pyncheon territory at the eastward, not engraved, but the handiwork of some skilful old draftsman, and grotesquely illuminated with pictures of Indians and wild beasts, among which was seen a lion; the natural history of the region being as little known as its geography, which was put down most fantastically awry. The other adornment was the portrait of old Colonel Pyncheon, at two-thirds length, representing the stern features of a Puritanic-looking person, in skull-cap, with a laced band and a grizzly beard; holding a Bible with one hand, and in the other uplifting an iron sword-hilt.[2]

A map and a portrait, geography and history. Hawthorne establishes these figures as the coordinates for the story, a framing device invoking past and future, as well as real and imagined. Earlier, the Colonel and the eastern territory had been mentioned, with both functioning as ongoing reminders of the nightmare of their past on the one hand and of the prospect for future happiness on the other. "Years and years after their claim had passed out of the public memory, the Pyncheons were accustomed to consult the Colonel's ancient map, which had been projected while Waldo County was still an unbroken wilderness."[3] The colonel "haunts" the family and the house; he is an avatar of the "tradition of all the dead generations" that Karl Marx says "weighs like a nightmare on the brain of the living."[4] The map, with its fantastic geography and natural history, represents hope, a promised land that may one day save the cursed family, if only it were realizable.

Of course, the novel's happy ending will come not from taking possession of the disputed land or by righting the ancient wrongs, but by making those fantastic things real, by evacuating their mysterious content in favor of plain reason and common sense. The "grotesquely illuminated" map, whose geography is put down "fantastically awry," is emended by the light of greater knowledge. The figure for such illumination is Phoebe, the girl who has the "proper gift of making things look real, rather than fantastic, within her sphere."[5] In this sphere, Maule's curse is lifted and the family feud ended through marriage; the burdensome past wastes away with the final death of Colonel Pyncheon (recast as Judge Jaffrey Pyncheon) and the discovery of the title deed to that mysterious territory to the east. Like other mysteries which are now explained, the terrible curse turns out to be a mere congenital health condition. That which was fantastic becomes altogether real.[6] In some senses, the resolution of the novel is figured in that imposing portrait and fantastic map; both history and geography are revised, corrected, and "put down" aright.

The geographical allegory implicit in *The House of the Seven Gables* requires no direct action on the part of any characters—no exploration of the eastern territory, for instance—in order to achieve its ends. By having the Pyncheon and Maule families' problems resolve themselves, as if by chance, Hawthorne allows the *representation* of the crisis to stand as the solution. That is, we are made to believe that, if only matters had been properly represented in the first place, none of this unpleasantness needed to have occurred. Once the imaginary geography is put down correctly, the other problems disappear. Although this view is typical of Hawthorne, whose characters often do not act so much as they are acted upon, Hawthorne is not unique. The association between narrative resolution and spatial or geographic understanding animates mid-nineteenth-century American literature, and the forces in motion during that era call for ever more nuanced forms of representation. In the United States, the experiment of democracy and the growing industrialization also met with rapid territorial and extraterritorial expansion, a dramatic rise in urbanization, and an unsettled and unsettling relationship between the personal, the local, the national, and the global. "The complex of capital, democracy, and industry seemed to tear the world apart, simultaneously unsettling from its place all that had existed and bringing into being ever fresh novelties."[7] This acceleration and intensification of social motion—perhaps most aptly figured in the newly emerged metropolis, which is both densely packed and disorientingly vast, or in the new technology of the railroad, which connects cities with hitherto unimaginable speed while also helping to disperse populations by enabling enhanced mobility—completely reshaped the imaginary communities and real spaces of nineteenth-century societies. The alienation or displacement caused by these movements called for new forms of writing and new ways to understand one's place in society and in the world. In the United States and elsewhere, the development of narrative forms was tied to the emergence of new social spaces. In other words, the new world needed new maps.

In exploring the spaces of narrative and the spaces of the emerging world system, Melville experiments with different narrative techniques and forms. In doing so, he develops a form of literary narrative to map the global space of the multi-, trans-, or supranational world system, but he also incorporates elements of national, personal, and local narratives as he struggles to comprehend the space of the nation-state, the extraterritorial contact zone of the Pacific, and the urban space of the metropolis. In subsequent chapters, I will examine these narrative forms and spaces of Melville's literary cartography of the world system, but here I want to explore the relationship between space and narrative form in nineteenth-century American literature.

"An actual locality to the imaginary events of this narrative"

Hawthorne's fear, in the preface to *The House of the Seven Gables*, that "The Reader may choose to assign an actual locality to the imaginary events in this narrative," goes beyond the mock horror of having his romance read as an actual exposé of the people and customs of "the County of Essex."[8] He is essentially carving a space for himself and his work that, despite all appearance, is removed from the actually existing Salem, Massachusetts. Yet his ironic defensiveness also highlights the degree to which his work, in the mid-nineteenth century, may be viewed as intentionally depicting social space it presents. The narrative cannot *but* stand in close relation to the geographic terrain, whether with respect to particular locales or to a more general geography. As D. H. Lawrence famously asserts, "Each continent has its own great spirit of place. Every people is polarized in some particular locality, which is home, the homeland. Different places on the face of the earth have different vital effluence, different vibration, different chemical exhalation, different polarity with different stars: call it what you like, but the spirit of place is a great reality."[9] The narrative forms of nineteenth-century American literature partake of this spirit in different ways.

As noted in the last chapter, the world system that emerged in the baroque era relies on the production of certain spaces, closely tied to the political and economic conditions and movements. Three distinct, yet closely related, social spaces or zones that emerge during the seventeenth century are also burgeoning in and around the United States in the nineteenth century: those of the state, or the space of the national; a peripheral zone, encompassing the extraterritorial domains towards which the state's centrifugal forces are directed; and the metropolis, the urban space related to the state's centripetal forces or power of centralization. A fourth space should be mentioned as well: that of the world system itself, a global space. As the world system develops, this global space becomes a challenge to the other spaces and to representation in and of itself. Effectively beyond representation, the global space must be figuratively drawn, often as a foil or backdrop to the other spaces. The forces in motion that enable a postnational world system to begin to emerge are often the same forces pulling apart the other social spaces. In the nineteenth-century United States, the challenge to produce an identifiable image of America is challenged and compromised by these other spaces and representational forms. At a historical moment when powerful social and economic forces are effectively shattering a comprehensive sense of place, the dominant drive in American literature is to achieve a sense of place that can be clearly understood as unique, distinct from the muddy complexity of the world: an ideal American place. In this way American literature develops as a kind of mapping project, or, rather, as a number of mapping projects. The narrative forms developed in nineteenth-century American literature functioned as engagement with these different spatial organizations. Narratives during this period sought not only to represent the spaces but also to transform them.

In *The Emergence of American Literary Narrative*, Jonathan Arac examines forms of prose narrative that emerge, thrive, and wane in American literature during the nineteenth century.[10] Adopting a generic approach to this literary history, Arac identifies and defines four distinct narrative forms: *national, local, personal,* and *literary*. These genres do not directly fit within other, more traditional, generic definitions. For example, in defining his terms, Arac makes little to no distinction between novel and romance, or even between fiction and nonfiction. And, although his chapters do emphasize certain decades over others, chronology alone does not determine whether a given work is included in a particular genre. Similarly, Arac does not specifically address extraliterary elements that are often used to characterize different genres, such as race, gender, or class; for the most part, men and women, white and black people, Brahmin and working-class writers may be engaged in producing any of the narrative forms. One thing that does underlie each genre, as I see it, is a particular spatial organization, a geographical frame of reference that does much to determine how a given narrative will fit within its formal parameters. Arac does not really state this himself, nor is it part of his intended project, but the geography of the narrative forms in nineteenth-century American literature is noteworthy. When the writers Arac examines elected to produce texts in the forms that they did, I maintain, they were also imagining the geographic and literary space in a certain, distinctive way. This spatial imagination relates directly to the form and to the content of their work.

I do not mean to suggest that antebellum American literature, its formal developments or its content, can be subordinated to a mapping project outside the literary endeavor. But writing has an inherently cartographic aspect, that in representing the social and physical spaces, writers (self-consciously or not) produce narrative maps that the reader may or may not use in forming his or her own spatial representations of the world. As Fredric Jameson has said of cognitive mapping, or the attempt to form a working image of the social totality in which one operates, it is "not merely that we ought to strive for it, but that we do so all the time anyway without being aware of the process."[11] To be sure, there are writers who self-consciously map their environs, right down to and including pictorial illustrations or hand-drawn maps in their pages. But in stressing the *literary* prong of literary cartography, I insist that the mapping project need not use actual diagrams or figures. The narratives are themselves maps, and the effects of their mapping are diverse.

It should also be mentioned that the formal analysis of literature does not preclude study of literature's historical or geographical aspects. The literary history of a nation is not the same as the history of that nation as seen through literature. Literature does indeed offer a vista into history, a point of entry for discussing the social, political, and cultural issues that the writings emerge from, respond to, or reflect. But literature is also an institution in its own right, with a history that may be related to other histories, but that cannot be reducible to them. "The central event in the literary history of mid-nineteenth-century American prose narrative is the emergence, around 1850, of works, pre-eminently *The Scarlet Letter* and *Moby-Dick*, that still count as 'literature' for many readers of the late twentieth century."[12] In the United States, the rise of literary works that would come to be valued as masterpieces of world literature occurred even as the country was struggling amid the turmoil of slavery, secession, continental and overseas expansion, problems associated with emergent industrialization and urban growth, and so on. Unquestionably, the writers of the antebellum period bore these matters in mind when they were writing, but many writers also made a conscious effort to develop and produce forms of literature that would operate in somewhat different registers, often far afield of these pressing concerns. In their very forms, narratives had specific functions, and the effects of narrative form can be far-reaching.

The literary narrative form, exemplified by *Moby-Dick*, Hawthorne's romances, and a few other works, is truly the anomaly of its era. Although twentieth-century literary studies tended to value such work above all others, most nineteenth-century readers and writers were far more interested in the kind of writing developed through the other narrative forms, in works that may cause problems for today's readers because the conceptual categories into which they fit are not always clear. National, local, and personal narratives do not comport with the rules and expectations associated with what is generally understood to be literary in its more specialized sense. At the beginning of the nineteenth century, the term *literature* referred to any culturally valued form of writing, and it would have certainly included a great deal of nonfiction—science writing, philosophy, travelogues, essays, history—in addition to poetry and prose fiction; indeed, in the sense that *literature* was used to designate "polite letters," novels themselves might have been excluded.[13] The nineteenth century, in Europe as in the United States, witnesses the transformation in the meaning of the word, and by the end of that century, one could frankly and matter-of-factly distinguish between the literary work produced by a Henry James in *The Ambassadors* and the nonliterary writing of, say, A. T. Mahon's *The Influence of Sea Power Upon History*, without much fear of debate. Both works could be valued, but for different reasons. Literature, unlike other forms of writing, was by then understood to be creative, imaginative, original, and so forth. But for most of the nineteenth century these qualities were not necessarily valued in themselves, and many writers, readers, and publishers preferred works engaged in other activities. Defined partly in contradistinction to national, local, and personal narratives, literary narrative in the antebellum United States was a more marginal form. The emergence of literary narrative was fraught with difficulties that ultimately rendered it a brief and evanescent moment in the literary history of the era.

The distinction between literary narrative and the other genres is not the same as that between fiction and nonfiction. While it is true that literary narratives were largely fictional, and that a type of imaginative, creative writing would come to define the form, the rival narrative forms frequently included (and sometimes combined) both fact and fiction. National narratives could be told through romances like *Uncle Tom's Cabin* or through the historical accounts like *The Conspiracy of the Pontiac*. Local narratives, following from Washington Irving's lead, included fanciful tales like "Rip Van Winkle" but also painterly, detailed sketches of actual places and people. Personal narratives, by and large, were nonfictional accounts of individual experiences, though many writers supplemented their narratives with fictional elements. The fact versus fiction divide is simply not determinative of the difference between literary narrative and the other forms. The difference lies in the form's relationship to the world.

"Those wonderful alterations which distinguish the progress of the American nation"

The dominant narrative form during the period covered by Arac's study is national narrative, which constructed an "America" often by narrating the story of the nation's colonial beginnings and then projecting its future as a world power. This story was could be told through fiction or nonfiction, and major figures working in this genre in the 1820s and 1830s included the novelist James Fenimore Cooper and the historian George Bancroft; later writers such as Harriet Beecher Stowe and Francis Parkman would also produce national narratives of great power. Arising alongside the profoundly patriotic rhetoric of Andrew Jackson's presidency, national narratives helped create a unified idea of the nation. At the time of their emergence and flourishing, there really was not a clearly national culture in the United States. "National narrative was part of the process by which the nation was forming itself and not merely a reflection of an accomplished fact."[14] Through national narrative, in part, the nation itself was formed.

The term *national narrative* is used slightly differently here than elsewhere in this book. Nineteenth-century national narrative of the type produced by James Fenimore Cooper is not exactly the same as that national narrative that forms the basis of twentieth-century American Studies, into which narrative literary works like *Moby-Dick, The Scarlet Letter*, or *Huckleberry Finn* get squeezed.[15] In the nineteenth century, national narratives often spoke directly to the pressing issues of the day, inviting if not demanding the reader to take a position on matters of national importance. By the twentieth century, thanks in part to the Myth and Symbol school of critics and other earlier practitioners of American Studies, the historically situated narrative became abstracted and made universal (thereby also, somewhat timeless). Thus, what Donald Pease calls the "image repertoire" of the American national community, connecting "an exceptional national subject (American Adam) with a representative national scene (Virgin Land) and an exemplary national mission (errand into the wilderness),"[16] establishes a template for determining the "American-ness" of a given narrative. As I discuss in Chapter 4, this imagery has been used to enfold *Moby-Dick* into American national narrative, but it must do so by ignoring much of the text's postnational form and substance. In the national narrative discovered, reconstructed, and embraced by American Studies, these mythic images allow for the abstract and iconic crystallization of the more fluid and muddied images produced in the nineteenth century. The national narratives of Cooper, Bancroft and others helped to create the image repertoire associated with Americanist national narrative, but the nineteenth-century versions stand as raw materials to the more finished product of twentieth-century critics, crude oil to refined gasoline.

Perhaps the greatest difference between the national narrative form in nineteenth-century literature and the national narrative of American Studies is that the former was created amid serious doubt that the nation itself existed in a unitary and definable form. The early practitioners of American Studies, many of whom were veterans of World War Two, could "believe in America," actively promoting the nation while also studying its features.[17] In exploring the myths of America, they discovered deeper truths, "the abiding values embedded in simple plots, the range and richness of formulaic metaphors."[18] But in Cooper's time, the simple plots and formulaic metaphors—as well as the abiding values—had to be invented.

The most salient images of nineteenth-century national narrative include the rugged individual, the democratic assembly, and the struggle to establish communities. It is easy to see how these translate into the image repertoire of Americanist national narrative, but it is also the case that the imagery is selective. In Cooper's work, for instance, a Natty Bumppo must be unique, not representative of the national community but always an outsider. His function is to help establish the community, but he does not settle down. Thus, in *The Last of the Mohicans,* Alice and Duncan (and their progeny) represent the Americans who will inevitably, as the fiction makes clear, control the continent, while Hawkeye (a.k.a. Natty Bumppo), whose expertise and daring made their happiness and secure future possible, exits stage left, along with the noble Indians (like Chingachgook) who must fade away and the ignoble Indians (like Magua) who must be killed. Although Natty Bumppo does conform to the archetypal image of the American Adam, he is more recognizably characterized in the Hollywood Western, a John Wayne or Clint Eastwood character, whose skill, knowledge and determination—combined with a clear lack of civilization and gentle manners—aids the settlers, protects the town, saves the damsels, and so on. In this way the space of the nation is expanded, while also allowing for the continuity that will make the "nation" recognizable.

Cooper himself draws the reader's attention to Natty Bumppo *as* a figure for the nation. In his 1831 Introduction to *The Last of the Mohicans,* he states,

> To portray an individual as a scout in the wars in which England and France contended for possession of the American continent, a hunter in that season of activity which so immediately succeeded the peace of 1783, and a lone trapper in the Prairies after the policy of the republic threw open those interminable wastes to the enterprise of the half wild beings who hang between society and the wilderness, is poetically to furnish a witness to the truth of the wonderful alterations which distinguish the progress of the American nation, to a degree that has been hitherto unknown, and to which hundreds of living men might equally speak. In this particular the fiction has no merit of invention.[19]

In other words, the narrative—and its protagonist—are truly representative of the American nation, despite what other romantic fictions may also appear in the text.

Despite its backwoods setting and antisocial woodsman, Cooper's *Leatherstocking Tales* depend on a version of the national that Bancroft's history exalts, a version based on the self-governing, small-group communities, embodied in the Puritan church congregation, the town meeting, and eventually the legislative assembly. Bancroft's *History of the United States,* begun in the 1830s, extended to ten volumes and was finally completed in 1874. Both Cooper and Bancroft were Democrats, supporters (if not entirely enthusiasts) of Andrew Jackson who identified with "the people" in their politics. In Bancroft's vision, the story of the United States, even before the revolution that transformed colonies into states to then be united, is the progressive story of ever-greater freedom, a narrative that retains tremendous power in political discourse, civics courses, and popular expressions of patriotism (or even critiques of it). The echoes are heard in the United States every Fourth of July, by every political candidate, regardless of party or position on matters of policy. By linking colonial congregations to post-revolutionary democracy, Bancroft is able to tell a unified, developmental story while also fostering an image of an eternal—that is, ahistorical—"America" that is present at all moments along the historical continuum Bancroft charts. As Arac notes, the paradox of this national narrative is that it relies on prolepsis, the anticipatory figure that sees "the United States" on the geopolitical maps of the early seventeenth century,[20] at the same time that it insists on progressive development. If the nation is already in place, why would one need to narrate its emergence and development? Of course, the answer lies in the mythic notion of "America," a figure that can always and unchangingly stand for freedom and equality, while also embodying an almost Hegelian sense of escalating, teleological progress.

Such a national narrative is clearly served better by establishing its origins: Puritans seeking religious freedom (as they saw it) in New England rather than greedy land barons in Virginia, rapacious Spanish conquistadors in Florida and the west, shrewd Dutch merchants in New York, and so on. Even in Bancroft's day, it must have been clear that the Puritan origin of the United States is historically inaccurate, even if the national narrative it provides is historically powerful and influential. As Sacvan Bercovitch has pointed out, citing America as "the example par excellence of collective fantasy,"

> we *know* that the Puritans did not found the United States. In fact, we know that by 1690, sixty years after the Great Migration and a century before independence, not even the colony of Massachusetts was Puritan. Nonetheless, the belief in America's Puritan *ur*-fathers was evident everywhere three centuries later, at every ritual occasion, from Thanksgiving Day to July Fourth, throughout literature, from Harriet Beecher Stowe to Thomas Pynchon, and in every form of literature, including endless debates about whether or not the Puritan legacy was a good thing.[21]

However, Bercovitch goes on to point out that, in another but also very real sense, America's origins do in fact lie with the Puritans, since the Puritans provide the rhetoric, the symbolism, and imagery of the collective fantasy that is the nation. National narrative thus creates and sustains the nation.

National narrative would remain the dominant American narrative form until the last quarter of the nineteenth century, but part of its appeal was that it provided an image of a national culture, and an imaginary national space, that could not yet be taken for granted. Although it may seem paradoxical to say that national narrative was dominant just as the nation as a political entity was about to be torn apart by secession and civil war, such crises actually explain the dominance of the narrative form. National narrative was so powerful precisely because the national identity was uncertain. It was not clear that the nation would form a unified whole, and national narrative's rhetoric of national unity and sense of purpose was often belied by the diversity and competing interests of its "people," some of whom did not readily identify with all the others. Since the nation was not a foregone conclusion yet, writers of national narrative could make a more or less compelling case for it. However, other narrative forms that flourished during these years reveal the extent to which the national space was not the principal frame of reference for many people in the United States. *Local narrative*, especially, exploited the notion of distinct regional voices, often presenting them to comic effect, but clearly undermining the idea of a homogeneous American people, unified in purpose and ideals.

"Little nooks of still water which border a rapid stream"

Washington Irving established the conventions that would define the local narrative form. Irving's detailed, often humorous sketches and his quirky Dutch characters provide foils to any idea of a representative national scene or subject. Rather than showing a microcosm of what America is, and should be, Irving presents a caricature. By presenting locales that are off the beaten track, removed from the mainstream of American culture, Irving self-consciously divides the local from the national. In a place like Sleepy Hollow, the

> population, manners, and customs remain fixed; while the great torrent of migration and improvement, which is making incessant changes in other parts of this restless country, sweeps by them unobserved. They are like those little nooks of still water which border a rapid stream, where we may see the straw and bubble riding quietly at anchor, or slowly revolving in their mimic harbor, undisturbed by the rush of the passing current.[22]

Indeed, "The Legend of Sleepy Hollow" could be read as an allegory, not for the nation, but in defiance of the national ideal. The only character in the tale who resembles an American Adam, a representative national subject, is Ichabod Crane, but Crane is ridiculed and run out of town by those very figures, the representatives of the old New York Dutch culture, who are doomed to fade away—or, what is the same thing, to be incorporated into the national culture—once the nation becomes fully formed. In the story's penultimate paragraph, we learn that Crane went on to study law, become a politician, and sit as a judge; in other words, he becomes an exemplary American public figure. Yet Katrina Van Tassel and Brom Bones are the nonrepresentative couple (unlike Cooper's Alice and Duncan), who presumably go on to sustain their own upstate subculture, rather than to embody a national culture.

The point of local narrative, after all, is not to present people with whom we, the readers, identify, but rather to display the oddballs. The reader is almost by definition an outsider, since the appeal of local narrative is based, in large part, on its presentation of an idiosyncratic elsewhere. The narrator in such tales functions as a native informant or sometimes as a bemused traveler who has stumbled across this strange subculture and who offers a glimpse of this local community to an outside world. Local narrative establishes the uniqueness of place, which tends to resist the homogenizing space of national narrative.

In the work of southwestern humorists like Augustus Baldwin Longstreet and George Washington Harris, the unique characters are clearly objects of ridicule. The reader is invited to laugh at them, and the tone is often condescending. Indeed, the humor is of an extraordinarily durable type, easily recognized in popular culture today. For example, a direct line may be traced from Longstreet's *Georgia Scenes*, offering tales and sketches of colorful, uneducated, and even barbaric denizens of that state, to the wildly popular "You might be a redneck" genre of jokes turned into a lucrative industry by Jeff Foxworthy, another "native Georgian."[23] Typically told from the point of view of an educated observer, southwestern humor depicted a way of life both foreign and risible to the readers. In the early nineteenth century, the "South-West" was still mostly east of the Mississippi River, and generally included Georgia, Tennessee, Kentucky, Alabama, Mississippi, and Arkansas; then, as now, these areas were considered somewhat backward, at least from the more urban (and urbane) points of view of polished city-dwellers in a Richmond, Charleston, or Savannah, not to mention Boston, New York, and Philadelphia. Certainly the backwoods rubes depicted in these stories were not meant to be representative of America, although the conventions would later be used by Mark Twain to produce more representative American types such as Huckleberry Finn. Indeed, as with other narrative forms, local narratives—particularly those of southwestern humor—could be incorporated into national narratives, as the once quirky backwoodsman is transformed into, or revealed to have been all along, another American Adam.[24]

More respectful in tone perhaps, but no less local in geographical and ideological scope, Hawthorne's short stories depict a way of life that also turns away from any national culture. His haunted Puritans, with their inward-looking brooding and secret sins, did not fit with the image of the rugged individual or the democratic community. Most often in Hawthorne's most memorable tales, the plot revolves around an individual whose personal ghosts prevent him from forming relationships with his fellows. Unlike the version of national narrative promulgated by Bancroft, in which ever-growing democratic communities interconnect to form that "America" that can transhistorically shine as a beacon to the world, and unlike Alexis de Tocqueville's sense that individualism itself unites the diverse people of the United States as "Americans," Hawthorne's tales emphasize the disruption of community and the deeply negative effects of individuality. In "Ethan Brand," for example, we learn that the Unpardonable Sin that Brand had searched for is actually produced by him and in him, when his heart "ceased to partake in the universal throb"; "He was no longer a brother-man," but "he was now a cold observer" of humanity, a "fiend."[25] Similarly, in "Young Goodman Brown," Hawthorne dramatizes the process of moving from community into isolation and disunity.

> Young "Brown," wedded to "Faith," seems intended as a generic, typical figure. Yet he proves capable of believing (he is not and need not be certain)—even heroically capable, insofar as the belief greatly pains and shocks him—that everyone except himself is secretly in league with the devil. The last paragraph of the story, however, in depicting the dreary end to which such suspicions doom Brown, makes an eccentric out of him. No longer typical, he is an isolated individual.[26]

For Hawthorne, the breakdown of communities, rather than their formation, often serves as the moral of the tale. His narratives thus are *local* in a more pointed sense than being merely regional. Hawthorne does offer painterly sketches of life in New England, but he also uses irony to show the localizing and alienating effects of individuality and the broken promise of community. In any event, a national culture does not emerge from Hawthorne's local narratives, notwithstanding his own use of the Puritan origins of American to reveal certain characteristics of the people.

If Hawthorne's haunted characters stand as foils to a representative national figure, how much more so do Edgar Allan Poe's? Poe's writings are anomalous in many ways, and his own canonicity is rather strange. For one thing, Poe's lasting popularity is at odds with his mixed reception among critics. Harold Bloom, in a volume I edited, calls Poe the New World's "worst" writer, whose "plain badness" is "palpable,"[27] and Poe was famously left out of Matthiessen's *American Renaissance*—partly because Poe died in 1849, a year before Matthiessen's key period (1850–1855), but also because his value "is now seen to consist in his influence rather than in the body of his own work [. . .] his stories, less harrowing on the nerves than they were, seem relatively factitious when contrasted with the moral depth of Hawthorne or Melville."[28] Poe, even more than Melville, seems *out of place* in American literature. In Arac's study, Poe is an anomaly as well. His work is included in the discussion of local, personal, and literary narrative—Poe's clearly produced no national narratives, and his hostility toward egalitarian democracy, among other things, prevented his works from being easily incorporated into national narrative later—but Poe's local narrative works do not partake of any particular region or place. His personal narrative (*The Narrative of Arthur Gordon Pym*) is extraordinarily unlike the others Arac discusses, and his literary narrative is based more in his literary theory than in any narratives he produced. Moreover, as Arac notes,

Poe is very difficult to understand, since his stolid and earnest tone contrasts with the unnatural or ridiculous content of the tales. "Readers were confident in laughing at southwestern humor and Hawthorne's complex ironies only emphasized his thoughtfulness, but Poe's work provoked fundamental uncertainty in response. Was he serious? Should his reader be serious? The problem is one of genre: What kind of work is this?"[29] Poe, like his characters, just doesn't seem to fit in America.

Although Poe's tales are not in any way regional, Arac does show how Poe's local narratives identify a *locale*, a particular type of space or place that is then explored through Poe's narrative mapping. Poe is perhaps the first American writer of the city, and Poe's career parallels the emergence and growth of cities in the United States. Not only does Poe's peripatetic career take him to nearly every major American city of his day—Boston, Charleston, Richmond, Baltimore, Philadelphia, and New York—but Poe's stories disclose an urban atmosphere, whether set in an actual metropolis (like "The Man of the Crowd" in London) or cloistered in some seemingly remote locale (as with "The Fall of the House of Usher"). Poe's dark and dreary settings resemble the cityscapes of Dickens and Balzac, revealing the power of atmosphere to influence plot and character. Unlike other writers of local narrative, Poe does not limit his geographic purview to a particular region, but his tales remain local in effect, as they carve out a small part of human experience and localize it in a particular space.

"There are some things in the narrative which will be sure to appear strange"

Whereas local narratives partake of place, "personal narratives arise from and depend on displacement."[30] The personal narrative form presents a first-person account of adventures in some exotic or foreign locale. In almost all cases, they are nonfiction (or based on nonfiction) accounts of the writer, but unlike the later modern autobiographies or the earlier Puritan narratives, these narratives rarely delve into the psyche of the narrator; rather, they serve as relatively impersonal reports from the periphery of polite society. Although the narrator does speak his or her own feelings, more often the narration deals with description of external events and places rather than with one's interior thoughts about them or lessons learned from them. Well known examples include Richard Henry Dana's *Two Years Before the Mast*, Melville's *Typee* and *Omoo*, Francis Parkman's *The Oregon Trail*, Frederick Douglass's 1845 *Narrative*, and Henry David Thoreau's *Walden*. Each of these works invites the reader to view the culture and character of populations of a 'low' social strata—working-class sailors, Polynesian or American natives, slave society, or the wilderness itself—that then becomes part of the reader's own world.

The readers of personal narratives, for the most part like the author, belong to a higher class and are members of the core-society. Except for the slave narrative, which generally moves upwards in an ascent from slavery to freedom, personal narratives normally follow a descent-and-return trajectory. That is, the writer/narrator—by choice (like Thoreau) or by chance (like Melville)—descends into the exotic space of the periphery, observes and records the experience there, and returns to metropolitan center, bringing not only the entertaining narrative but an almost ethnographic study of the foreign places. Although the experiences are, by definition, foreign to the national narrative, personal narratives are easily incorporated into national narrative. "A generic appeal of personal narratives in their time and since is their registration of what seems a more archaic way of life, a virtual past achieved by travel in space rather than in time, but from the perspective of a narrator who is, like the readership, part of a modern world, making contact with the 'other' world and transforming it while integrating it. Personal narratives may act thereby to colonize places and kinds of experience, which are then appropriated into national narrative."[31]

The authority of personal narratives rested on their being true. As such, they were valued not for their originality and imaginative power, but mainly for their authenticity in reporting the experiences. A chief criticism of personal narratives, as was the case with the controversy over Melville's *Typee* (which I discuss in Chapter 6), was that the story could not be true. In fact, the publishing house Harper and Brothers turned down the manuscript for *Typee* on the grounds that "it was impossible that it could be true and was therefore without value."[32] Poe's utterly fantastic *Narrative of Arthur Gordon Pym of Nantucket* was nevertheless published (by Harper and Brothers, no less) as nonfiction, and continued to be advertised as a work of travel for years.

Because of their value as truthful reports from an exotic zone, these narratives often had real effects in social and political discourse. As with Douglass's, personal narratives were often overtly political documents designed to persuade readers to adopt a particular position on matters of significance. Similarly, both Dana and Melville criticized the practice of flogging in both civilian and military ships, and their personal narratives were influential in debates on the subject. The value of personal narratives lay in their ability to provide accurate reports, which could then be used to effect real reforms. The space of a personal narrative—itself a map of a peripheral, perhaps extraterritorial zone—thus is incorporated into a spatiopolitical system, both in understanding the foreign spaces and in transforming them, along with the more familiar national and local spaces.

The dialectic of the strange and the familiar in personal narratives assured their success, as readers hungered for the exotic; the requirement that the narratives be true convinced the audience of their inherent value. Melville acknowledges as much in the Preface to *Typee,* in which he tempts the reader by conceding that the narrative might be strange, even unbelievable, while also being entirely nonfictional.

> There are some things related in the narrative which will be sure to appear strange, or perhaps entirely incomprehensible, to the reader; but they cannot appear more so to him than they did to the author at the time. He has stated such matters just as they occurred, and leaves every one to form his own opinion concerning them; trusting that his anxious desire to speak the unvarnished truth will gain for him the confidence of his readers.[33]

The strange and incomprehensible extraterritorial places are thus enfolded into a map of the familiar or homey through the personal narrative. Yet the idea that narrative might project a world beyond the familiar map was, in the mind of those demanding the personal narrative form, still a stretch.

"So like the real world"

As noted above, Hawthorne complained that the demands of certain readers were impinging on his rights as a "romancer." Specifically, he demanded the latitude to create a world in the fashion entire of his own. In the preface to *The Blithedale Romance*, he argues that the United States itself was part of the problem, since unlike European countries it had not developed a tradition of fiction that afforded the author such freedom. In the U.S., Hawthorne complained, "there is as yet no such Faery Land, so like the real world, that, in a suitable remoteness, one cannot tell the difference, but with an atmosphere of strange enchantment, beheld through which the inhabitants have a propriety of their own."[34] For Hawthorne, literary narrative should be able to create a space outside of the national, local, and extraterritorial spaces—a literary space of the imagination. For a brief moment in the middle of the century, Hawthorne's "remote" zone that was "so like the real world" was possible.

Arac argues that the Compromise of 1850 opened a narrow window for literary narrative to flourish, visible in *The Scarlet Letter* and *Moby-Dick*.[35] Marked by a decidedly ambivalent attitude toward real-world concerns, literary narrative created a space outside of the national culture and populated it with imaginary people and places. Drawing from the Romantic tradition, these works celebrated individual genius, creativity, originality, and novelty; this, in turn, involved a reappraisal of the common, familiar, or popular as unworthy of true artistic achievement. Explicitly in Poe, and more implicitly in the democratic Hawthorne and Melville, literary narrative suggested that a specialized audience was needed to understand or appreciate it. If novelty or originality is a key attribute, then a great literary work requires not only a gifted author, but a skilled, well qualified reader who can fully appreciate the genius of literary art. That is, it requires a reader who can discern ingenious novelty from mere poorly executed examples of older forms. Almost by definition, then, the finest literary texts must be unpopular, since the masses are unlikely to recognize bold new art (which might look, to the untrained eye, like failed attempts at the older, familiar styles). Thus Poe could claim that, while Hawthorne "evinces extraordinary genius, having no rival either in American or elsewhere," his favor among literary critics has had "little effect on the popular taste"; Hawthorne, Poe asserts, is "*the* example, *par excellence*, in this country, of the privately admired and publicly unappreciated man of genius."[36] Similarly, Hawthorne, in lamenting his own meager sales in a letter to his publisher, impugned the work of more popular writers of the day; most famously, he derided the "damned mob of scribbling women," saying "I should have no hope for success while the public taste is occupied with their trash."[37] And Melville could complain to Hawthorne melodramatically, "Though I wrote the Gospels in this century, I should die in the gutter."[38] All of this speaks to bitterness, perhaps jealousy, but it also registers a distinct break between what these authors feel is great literature and that which can achieve popularity. For Washington Irving or James Fenimore Cooper, both quite popular and thoroughly admired, such a distinction was not pressing. The writers of literary narrative felt distant and, in their work, actively distanced themselves from the public. The writer, not unlike the characters found in Poe, Hawthorne, and Melville, is alienated from society; like Ethan Brand, he has become an observer of humanity, and is no longer a member.

This distance played out in other areas as well. Whereas a writer of national narrative was interested in speaking to issues of popular and national importance, the writer of literary narrative strove to deal in universals, abstract ideas often far removed from the muddy realities of social or political discourse. As I discuss at greater length in Chapter 6, literary narratives had a different relationship to "the truth." Unlike personal narratives, which demanded to be taken as true and which were even accompanied by evidentiary authority and vouched for by witnesses, writers of literary narratives wanted to be left alone to create their own imaginary worlds. In his preface to *The House of the Seven Gables*, Hawthorne writes that the romancer "wishes to claim a certain latitude, both as to its fashion and material," and that the romance has the right to present its truth "under circumstances, to a great extent, of the writer's own choosing or creation."[39] Hawthorne self-consciously turned away from the social world to create a freely imaginative space. Literary narrative could transcend, and thereby implicitly or explicitly criticize, the world of everyday experience. But by transcending the everyday, it also moved outside the sphere of the common man, becoming an elitist enterprise, requiring a special kind of reader as well as a special kind of writer. Once the writer became an artist, the relationship between the writer and the public forever changed.

Literary narrative, as theorized and practiced by Poe, Hawthorne, and Melville, courted subtlety, ambiguity, complexity, even obscurity. In this sense, literary narrative is *baroque*. These authors were most interested in exploring the psychology of their characters, presenting abstract problems for which no solutions readily presented themselves and leaving questions open for the reader to contemplate. It is not that these narratives refused to speak to local or national concerns outright, but they insisted on doing so in a language that many readers would have found coded or indecipherable. Whereas a national narrative like *Uncle Tom's Cabin* could offer a clear point of view, expressing a definite position on a matter of national urgency, and exhorting the reader to take sides, a *Moby-Dick* would offer a masked narrator ("Call me Ishmael"), brooding about matters, even pressing political ones, in such a way as to inspire only more questioning. For example, the perceptive rhetorical question Ishmael poses in connection to his need for money and to his submission the authority of others—"Who ain't a slave?"[40]—might elicit thoughtful contemplation of the state of man in a postlapsarian condition or in a capitalist economy, but it hardly speaks to the issue of actual chattel slavery, a matter much on the minds of readers in the United States in 1851. It is not that literary narratives avoided real-world politics, but that they insisted on presenting politics in abstract, often ambiguous ways.

Not in spite of, but precisely because of its imaginative power, literary narrative could not survive the crisis of an impending Civil War. In the years just after *Moby-Dick*, the powerful national narrative *Uncle Tom's Cabin*, with its straightforward, first-person narrator holding strong and unambiguous views on issues of national importance, would become a best-seller that would shake up the world, both the literary world as well as the realms beyond *belles lettres*. *Uncle Tom's Cabin* left no doubts where it stood or where it wanted its readers to stand. In the same year that Melville puzzled readers with the "ambiguities" of *Pierre*, Stowe enflamed the passions of abolitionists and stirred the anger of proponents of slavery. As Arac notes, "literary narrative could not maintain its separate realm against the crises that from *Uncle Tom's Cabin* to the end of Reconstruction again brought to the fore national narrative."[41] In the wake of the Civil War, once the nation-state had been reformed and its power consolidated, literary narrative reemerged and, by the turn of the twentieth century, triumphed. National narrative, as suggested above, waned once the nation and the power of the nation-state were no longer matters of doubt. With the advent of modernism, the "literary" achieved far greater status, and national narrative became relegated to nonliterary discourses (notably popular culture and politicking). Literary narrative, like the baroque, returned in modernism to become a model for new generations of artists attempting to map their new worlds.

The four narrative forms discussed in this chapter correspond to four spaces, or rather to four spatiopolitical sets in which various spatial relations are in motion at once. These spaces, like spaces on a map, are imaginary and real at the same time, affecting the way one thinks about oneself vis-à-vis the world and affecting one's activities in the world.[42] The space of the nation, which is shaped and mapped by national narrative, dominates the discourse of the nineteenth century, and has become a defining aspect of twentieth-century cultural studies, notwithstanding the waning of the power of the nation-state in the post-World War Two era. The extraterritorial zones of the periphery, which are mapped by personal narratives, can be characterized as exotic spaces on the one hand, and utopian spaces on the other. Both become important to the space of national narrative, as the "no place" and the authentically "true places" of the periphery are incorporated into a national ideal. The idiosyncratic space, and especially the urban space, of local narrative manifests those power that trouble national narrative and its mapping. Eventually, the space of the urban undermines the national by connecting the intensely local experience to a global or supranational world-system.

In subsequent chapters, I examine these overlapping territories and narratives in the context of Melville's literary cartography. In Chapters 4 and 5, which provide a reading of *Moby-Dick* against the grain of nineteenth-century national narrative and of twentieth-century American Studies, I argue that Melville's work is already postnational, and that it attempts to map an emergent world system in which the nation-state no longer embodies the central power or even plays the leading role. In Chapter 6, I look at Melville's South Seas cartography and his use, then critical rejection, of the personal narrative form. In his critique of personal narrative, Melville embraces a more global form of literary narrative while also rejecting nationalist discourse, and he places the geographical periphery in relation, not to the core-societies *per se*, but to the world as a whole. In Chapters 7 and 8, I look at how Melville's critique of personal narrative is finally driven by the seemingly unmappable *locale* of the city, the ambiguous urban space that thwarts Melville's attempted map and leads to a reconsideration of the mapping project. The metropolis becomes a problem for representation, and a key site of the world system requiring new understandings and techniques. Melville's literary cartography grapples with the interrelations of national, personal, and local narratives and the spaces with which they are associated respectively.

A fourth space, associated with literary narrative, is the space of the world system itself, a global or planetary space.[43] In literary narrative's attempt to project a world elsewhere, as Melville employs the form, it becomes a sense of mapping the world itself. This world partakes of the national, personal, and local spaces, but it also transcends and subsumes them, like a Hegelian *Aufhebung* without the unidirectional teleology. The space of the global is thus not merely another space added to these others, but a defining space, a horizon beyond which these others cannot really be imagined but which does not always come into view. The baroque literary form of Melville's work is well suited to mapping this space, as the form and content of a book like *Moby-Dick* indicate that it is a *world text*, containing yet projecting beyond other narrative forms and spaces.

Chapter 3

"An everlasting terra incognita": Globalization and World Literature

When Wellingborough Redburn arrives in Liverpool, a major seaport in the most powerful maritime economy in the nineteenth-century world, he marvels at the multinational diversity on display at the docks. In language not unlike Melville's description of New York City's wharves at the beginning of *Moby-Dick,* Redburn notes how the world market, embodied in circulation at sea, has transformed and laid bare the pretensions of nationality itself.

> Surrounded by its broad belt of masonry, each Liverpool dock is a walled town, full of life and commotion; or rather, it is a small archipelago, an epitome of the world, where all the nations of Christendom, and even those of Heathendom, are represented. For, in itself, each ship is an island, a floating colony of the tribe to which it belongs.
>
> Here are brought together the remotest limits of the earth; and in the collective spars and timbers of these ships, all the forests of the globe are represented, as in a grand parliament of masts. Canada and New Zealand send their pines; America her live oak; India her teak; Norway her spruce; and the Right Honorable Mahogany, member for Honduras and Campeachy, is seen at his post by the wheel. Here, under the beneficent sway of the Genius of Commerce, all climes and countries embrace; and yard-arm touches yard-arm in brotherly love.[1]

In Redburn's view, the Liverpool docks themselves form an "Anacharsis Clootz deputation from all isles of the sea, and all ends of the earth,"[2] a figure for internationalism and even postnationalism. Indeed, in this chapter on the diversity of nations represented at the docks, Melville utters his most famous comment about the specifically *American* nationality. More to the point, he rejects that there is such a thing, arguing that the national identity of the United States can only be viewed in its fundamentally multinational character: "We are not a nation, so much as a world; for unless we claim the world for our sire [. . .], we are without father or mother."[3]

It is thus perhaps trebly odd, then, that Melville would become so representative an author for a thoroughly nationalistic American Studies and that *Moby-Dick* would become one of the most sacred texts in that well nigh religious field of study. I say "trebly" odd, because there are at least three particular aspects of Melville's work that resist the founding premise of American Studies. First, in the matter of the *content* of American national narrative, Melville denounces the notion of an Adamic individual subject traversing a Virgin Land on a divinely ordained errand into the wilderness. These key terms, forming the image repertoire of American national narrative, are, if anything, savaged in *Moby-Dick*. The novel satirizes these images even as it presents a clear post-national world system, with an expressly non-Adamic collectivity ensconced in a worldwide industry (hardly "virgin" soil) pursuing a commercial enterprise of commodity production in a moment of intense historical, political, and economic crisis. Far from presenting any representative American individual, scene, or activity, Melville labors to produce an image of the global space of the world system and the elements which make it up. Melville, in all of his works but especially in *Moby-Dick*, offers a profound critique of nationalism while projecting an image of a postnational world system in its emergence.

Second, at the level of *form*, Melville subverts the familiar guidelines and expectations of national narrative in his attempt to produce a distinctly literary narrative, to continue to use Jonathan Arac's terms for the time being. It is not merely that his work ironically undermines the content of national narrative, but Melville presents his postnational narrative in a baroque form that serves to estrange rather than to make familiar. The formal extravagance can be identified in the use of a distancing narrative voice or voices, a bizarre concatenation of stylistic effects that alienate the reader from the author, and the incorporation of multiple forms in order to effect something like a Hegelian *Aufhebung* of the various modes of popular writing of the day. Melville cancels, preserves, and elevates or sublates the other forms even as he constructs his own baroque text.

A third aspect, incorporating both form and content, involves the *theory of the novel* itself. Most of Melville's work, but especially *Moby-Dick*, does not adhere to the formal guidelines of national narrative, and Melville tends to undermine the expectations of the novel or romance in general. Never mind that several generations of Melville scholars, from the initial Melville Revival of the 1920s to the academic canonization of Melville in the 1950s and on to the revisionist treatments of "New Americanists" in the 1980s and 1990s, have essentially viewed *Moby-Dick* as the representative novel (along with *The Scarlet Letter*) of antebellum American literature and as the representative text of an American Renaissance. But it might be noted that *Moby-Dick* is not merely an eccentric novel, an extravagant romance, or an odd narrative; it is not really a novel at all (at least, not in a traditional view of the genre). As I argued in Chapter 1, *Moby-Dick* is a baroque text, a bizarrely extravagant encyclopedia, containing or attempting to contain the entire world. It certainly does not operate as an expression of the nation itself, although *Moby-Dick*'s "outreaching comprehensiveness of sweep" would include many "nations and generations" in its ambitious cartography. As William Spengemann has suggested in reference to *Pierre*, "Melville didn't write novels." Moreover, Spengemann asserts, "With certain notable exceptions, he didn't even read novels."[4] The novel form, however else it might be defined, tended in the nineteenth century to become the representative literary form of the nation-state itself, both in its widespread popularity and its ability to create an "imagined community."[5] With *Moby-Dick*, the national scope associated with the novel is far exceeded, and the baroque production indicates something much greater: an entire world system. This is partly what Franco Moretti means when he calls *Moby-Dick* (among other "modern epics") a *world text*.[6]

Melville is much more a "global" writer than an American one, and *Moby-Dick* is a world text, not a national one. As I argue throughout this book, Melville's attempt to map a world system in its emergence involves a critique of nationalism and nationalist forms, while it also suggests new ways of understanding the world as a whole. Part of that may be seen in Melville's interests in circulation and exchange, in the international and transnational movement of persons, goods, and forces (as Redburn witnesses at the Liverpool docks), and in the power of literary narrative to give form to, or make sense of, a world system too vast to "know" in traditional ways. This world, like the vast ocean, is "an everlasting terra incognita."[7]

World Market and Literature

The idea of *world literature* is almost always associated with Goethe, who named the term (*Weltliteratur*) and gave it force in a number of writings in the early nineteenth century. Strikingly, for instance, at a time when many national literatures—including German literature, of course—were only beginning to achieve the esteem and recognition that nationalists might have hoped for, Goethe asserted that "National literature is no longer of any importance; it is time for world literature, and all must aid in bringing it about."[8] What Goethe meant by *Weltliteratur* varies at different times and in different contexts. Sometimes, as in the quotation above, he seems to call for a postnational literature, one that transcends and renders obsolete any national literatures that preceded it. At other times, Goethe seems to call for a kind of United Nations of Literature, where writers and critics of different national cultures may come together to compare ideas, share their work, and promote literature generally. In another version, Goethe sees world literature as the culmination or apotheosis of national identities, the essence of individual nations projected onto a world stage. Elsewhere, world literature simply registers the degree to which the idiosyncrasies of individual nations may be recognized as minor, superficial characteristics in a world increasingly brought together by enhanced means of communication and circulation. In all of these variations, notably, elements of what came to be understood in the late twentieth century as *globalization* played their part, whether that be understood in terms of communication technology, transportation, international political relations, cultural cross-pollination and influence, economic developments, liberal trade policy, new forms of finance, and so on. In other words, the development of a socio-economic world system entails the development of world literature.

In *The Communist Manifesto,* Marx and Engels make this connection between the world market and literature explicit, while also recognizing "the national" as an essentially reactionary or backward form.

> The bourgeoisie has through its exploitation of the world market given a cosmopolitan character to production and consumption in every country. To the great chagrin of Reactionists, it has drawn from under the feet of industry the national ground on which it stood. All old-established national industries have been destroyed or are daily being destroyed. They are dislodged by new industries, whose introduction becomes a life and death question for all civilized nations, by industries that no longer work up indigenous raw material, but raw material drawn from the remotest zones; industries whose products are consumed, not only at home, but in every quarter of the globe. In place of the old wants, satisfied by the production of the country, we find new wants, requiring for their satisfaction the products of distant lands and climes. In place of the old local and national seclusion and self-sufficiency, we have intercourse in every direction, universal inter-dependence of nations. And as in material, so also in intellectual production. The intellectual creations of individual nations become common property. National one-sidedness and narrow-mindedness become more and more impossible, and from the numerous national and local literatures, there arises a world literature.[9]

Weltliteratur, in other words, is part and parcel of the dynamic processes of the establishment and spread of the *Weltmarkt.*

The finality with which Goethe and Marx pronounce the demise of national literatures and the assured emergence of world literature exposes their considerable gift of hyperbole. By the mid-nineteenth century, neither world literature nor the world market could really be said to have been firmly established, although it was clear to Goethe, Marx, Melville and others that the tendency toward the global, in both the economic sphere and the cultural, was ineluctable. As indicated by Marx and Engels in the passage above, the resistance to this nascent and emergent globalization was largely driven by reactionary politics, by a historically right-wing and retrogressive view of culture, history, and economics. In other words, at a time of intense nationalism and a consolidation of state power in Europe, Marx could recognize national culture as an already *residual* formation in the face of an increasingly global political economy.

In a fascinating fragment from the notebooks later collected and published as the *Grundrisse,* Marx criticizes the political economic theories of Henry Carey, whom Marx calls "the only original economist among the North Americans."[10] In their classical liberalism, many of Carey's ideas are eerily similar to those prominent in the discourse of neoliberalism in the late twentieth and early twenty-first centuries, especially with regard to free trade, expanding markets, limiting taxation and tariffs, and globalization. And the same paradoxes are present. Carey, not unlike those neoliberals who call for a "downsizing" of government and a more *laissez faire* approach to economic regulation, believes that state interference is the principle obstacle to economic prosperity. His model is England, to whose economic relations he opposes those of America. Once the market is free from state influences, Carey seems to think, economic harmony is inevitable. "To what extent these State influences (public debt, taxes, etc.) themselves arise from bourgeois relations," Marx notes, "and to what extent even in North America the power of the central government grows with the centralization of capital—this, naturally, Carey does not investigate."[11] But this is not the central paradox in Carey's theory. The paradox arises when Carey is faced with the awesome reality of the world market. International competition, and especially England's dominant position in this competition, is for Carey the greatest threat to domestic economic harmony. Marx observes Carey's position with irony:

> The only defense against this [world market competition] are protective tariffs—the forcible isolation of the nation from the destructive power of English large-scale industry. Therefore the State, branded at the outset as the only disturber of the "*harmonies économiques,*" becomes their last refuge [. . .]
>
> [T]he harmony of bourgeois relations of production ends with Carey in the total disharmony of these relations just where they appear on upon the most magnificent scene, the world market, and in their most magnificent development, as the relations of producing nations. All the relations which appear to him harmonious within particular national boundaries, or also in the abstract form of general relations of bourgeois society—the concentration of capital, division of labour, wage labour, etc.—appear to him as disharmonious where they show themselves in their most developed form—in their world market form—as the internal relations which produce England's domination in the world market and which, as destructive influences, are the result of this domination [. . .]
>
> Carey has not grasped that these world market disharmonies are only the ultimate adequate expression of the disharmonies which have become fixed in the economic categories as abstract relations or have a local existence on the smallest scale.[12]

Carey's attitude towards the state is not only paradoxical, Marx suggests, but also fatally flawed, insofar as that sort of protectionism cannot succeed in an era of emergent, multinational capitalism. So pervasive, even in the mid-nineteenth century, is the world market, and by extension the emerging geopolitical world system, that neither "pure" economic relations (e.g. on a "local" scale) nor the state itself can escape its influence.

Moby-Dick, Melville's most conspicuously baroque text, emerges amid the crises which occasion these paradoxes, calling into question the role of the nation-state at a moment when, simultaneously, the state is consolidating its central power and it is threatened by internal and external forces. This moment also sees the organization in America of a nationalism linked to territorial expansion, expressed not only through the official channels of political rhetoric and economic policy (e.g. Carey's ironic protectionism in support of free enterprise), but also through developments in national narrative in the burgeoning American literary scene. *Moby-Dick* responds to these circumstances by projecting a figure of the multinational world system, which in turn functions as a critique of the nationalist and statist forces in literature and society. That is, *Moby-Dick* presents a cartography of an emerging world system in which the nation-state and its apologists and ideologues are subsumed by the forces of this supranational Leviathan.

Of course, *Moby-Dick* does not simply dismiss the power of the state and of nationalist drives; rather, the novel combines an intense consideration of these powers with a projection of a world system that supersedes and incorporates them. By my reading, *Moby-Dick* begins with a philological exposition in its first two, unnumbered chapters, following a tradition of thinkers directly interested in the problems of nation-formation and language (notably, Vico and Herder, whose influence on Goethe and hence on the formation of the idea of *Weltliteratur* is significant). The study of language, and especially of written texts, develops out of the will to identify, to understand, and eventually to construct the nation. From this tradition unfold the linguistic and cultural foundations of national narrative. For example, as Isaiah Berlin points out, while Herder is the principle exponent of the historical views of national formations, "the man who first gave it flesh and blood was Sir Walter Scott."[13] And Scott's historical novels became the primary literary influence on those writers (especially James Fenimore Cooper) who were establishing an American national narrative in the United States. *Moby-Dick*'s philological beginnings, then, offer a point of departure for the novel which will engage in a meditation on the national while also projecting an image of a postnational world system.

Space and Literary Narrative

As I argued in the last chapter, one may assign a kind of topographical correlation between the narrative forms in nineteenth-century American literature and the social spaces of the United States. That is, each narrative form has a distinct *topos*, a spatiopolitical set of relations understood to be associated with a given way of imagining one's relations with the world. For national narrative, the space of the nation-state is both projected and assumed, in a dialectical play of representation and presentation, or—in the German critical tradition of Kant, Hegel, and Marx—of *Vorstellung* and *Darstellung*. In describing a distinctly American nation space, national narrative produced its own imaginary geography, and established the particular space that other narrative forms and their corresponding *topoi* would be defined against. The local narrative form, in its emphasis on regional specificity and in depicting the most visibly distinctive features of the locale that hindered the assimilation of local places into the national space, carved out domains different from, and often at odds with, the idealized national geography. Not only did local narratives distinguish *places* from the national space, but the narrative form highlighted the different *kinds of place*. That is, it is not merely that the rural South is different from the increasingly industrial Northeast, but that there are profound differences even within a region between different social spaces—most famously, in an era of rapid urbanization, between the country and the city. The local narratives of Poe, which influenced Melville's own tales ("Bartleby, the Scrivener," for example), are frequently imbued with an urban atmosphere, even when set outside any particular city, and they drew out the spectral, eldritch, or startlingly unfamiliar character of the new spaces that, while located within the space of the nation, are also foreign to it. With personal narratives, the extraterritorial or exotic spaces outside of the national domain are mapped and made familiar as they are incorporated into the broader national image. These spaces thus have the dual function of introducing the foreign element while extending the national imaginary space. As such, personal narrative begins by mapping contrasting spaces (exotic versus familiar) and ends by enfolding the one into the other.

Literary narrative sought to create and to represent a different kind of space altogether, but a space that also encompassed the others. Its authors often self-consciously strove to conjure up a space of the imagination, a roadside theater (in Hawthorne's image) off the beaten path of quotidian reality. In seeking a space of imaginative freedom, of artistic innovation and originality, the writers of distinctly literary narrative seemed to turn away from the pressing concerns of local, personal, and national life. In some cases, this literary narrative turns inward, focusing on psychological development and character, on private moral questions. For Arac, the turn to literary narrative involved a decisive movement away from the real world and into a realm of imagination, occasioned and enabled by the Compromise of 1850 that ever so briefly allowed the most pressing national controversies to be placed on the back burner. Not surprisingly, the moment for literary narrative in the antebellum United States is just that, a *moment*. With the appearance of *The Scarlet Letter* in 1950 and *Moby-Dick* (and *The House of the Seven Gables*) in 1851, literary narrative leaps onto the scene in Arac's history of narrative forms. But in 1852, the year of *The Blithedale Romance,* Melville's catastrophic *Pierre* exposed the limits of literary narrative in the United States in that era. The same year, a powerful national narrative in the form of a novel, *Uncle Tom's Cabin,* emerged and transformed both the literary marketplace (by becoming an enormous "bestseller") and the national discourse in the public sphere. Literary narrative, ostensibly removed from the concerns of most readers and favoring complexity, even ambiguity, over clearly articulated political positions, was not able to hold most readers' interest in the face of such epochal crises as the impending Civil War. The national space, a political geography that was in real danger of being thorough redrawn, overwhelms the spaces of literary narrative.

Although I agree with Arac that "literary narrative could not maintain its separate realm against the crises that from *Uncle Tom's Cabin* through the end of Reconstruction again brought to the fore national narrative"[14] in the United States, I argue that literary narrative, as imagined and produced by Melville in particular, did not so much seek to establish a "separate realm" as to project a more comprehensive space that would include all these others. For Melville, literary narrative attempts to represent a totality of experience, a totality that other forms could not (and whose writers were not interested in trying) to represent. As I discuss at greater length in Chapters 6 and 7, Melville abandoned the narrative form with which he had had such commercial and critical success, the personal narrative, in favor of a form that offered him the best means of producing a truer map, a representative form that would encompass, or attempt to encompass, the global totality. For Melville, *literary narrative* was not merely a form used to escape from the cramped confines of local, personal, or national spaces, but it was the form for representing global spaces well beyond them.

Although Hawthorne's romances, with their private-sphere turmoils and their soul-searching or angst-ridden characters, might serve as a model for mid-nineteenth-century literary narrative, the global reach (if not quite global grasp) of the form might be seen best in the first two literary narratives Arac identifies, at least chronologically: Poe's *Eureka* and Melville's *Mardi,* each appearing in 1849, one year before *The Scarlet Letter.* Poe's elaborate, cosmological prose poem literally, if somewhat satirically, presents the entire universe, in its intensity and diversity as well as in its "oneness." Melville's South Seas "romance," in which he deliberately distanced himself from his earlier personal narratives, quickly reveals itself to be a global narrative; the chapter titled "World Ho!" announces that the novel will embark upon an actual circumnavigation of the world while also endeavoring to touch on all matters worthy of discussion—a world and also a world of ideas. In both texts, a representation of the global totality is the goal. Indeed, in Poe's literary theory, what Arac calls "literary narrative" derives from the successful presentation of "totality."[15] In Melville's own case, the need to map a totality (as Fredric

Jameson has phrased it) comes across in his anxiety over omission. As he quotes Hackluyt on the very first page of *Moby-Dick*, Melville indicates that when you leave out an essential element, "you deliver that which is not true." Melville's concern for delivering that which is true, paradoxically perhaps, led him away from the largely nonfictional personal narratives and towards the much more overtly fanciful literary narratives of *Mardi*, *Moby-Dick*, and *Pierre*, among others. Literary narrative, for Melville especially, is designed to map global literary spaces.

In *Moby-Dick*, it is not merely the "circumnavigating Pequod" or the multinational workforce of "an Anacharsis Clootz deputation from all the isles of the sea, and all the ends of the earth" that make for the global narrative. It is also the sentiment of the Hackluyt quotation and the "outreaching comprehensiveness of sweep." The novel's *Ansatzpunkt*, its beginning or jumping-off place, in "Etymology" and "Extracts" indicates the ambitious, multinational scope of *Moby-Dick*, as multiple languages and discourses will serve to frame the narrative; reference is here made to "all the known nations of the world" and "many nations and generations."[16] *Moby-Dick* goes so far as to include the other narrative forms Arac discusses, forming its own compendium of narrative forms even as it establishes its own baroque form. The "Call me Ishmael" beginning to "Loomings" signals a kind of personal narrative, of course, one which Melville will quickly subvert. (In the next chapter, I discuss how a canonical interpretation of *Moby-Dick* has reinstated Ishmael as the author of a personal narrative, only then to transform the novel into a national narrative, or more aptly, to effect the nationalization of the literary narrative.) The images in the first hundred pages of *Moby-Dick* evoke those of local narratives or sketches, as the hypochondriac Ishmael and the quirky New England whaling communities resemble elements from "Down East" humor about Yankees; the grandiose hunt for the legendary white whale recalls the "Big Bear of Arkansas" or other tales of southwestern humor. "Yet the immensely greater size of *Moby-Dick* breaks the generic mold, and its resemblance to two different kinds of local narrative suggests that it can properly belong to neither."[17] And obviously, the rhetoric of national narrative, from its fierce advocacy of the principle of equality to its many references to patriotic images, *Moby-Dick* contains—in multiple sense of the word—national narrative. However, I will add that much of this rhetoric is twisted back upon the purveyors of national narrative, as when Melville insists that "we are not a nation so much as a world" or "Queequeg was George Washington cannibalistically developed." Melville's seeming embrace of the national almost always quickly shifts to the forthright promotion of a transnational or postnational ideal. The variety of narrative forms in *Moby-Dick* comports with its baroque character and with Melville's view that the "true" literary narrative must not leave anything out.

Melville's literary narrative charts a global literary space.[18] Its scope is the world itself, and both its form and its content strive to map the global totality as well as Melville could manage. In its baroque excess, *Moby-Dick* also wanders out of bounds—the very definition of *extravagance*—not only borrowing from various genres and styles, but utterly transforming them. In many respects, as noted above, *Moby-Dick* is not a novel, but something else, an unclassifiable hodgepodge of other genres or, as Evert Duyckinck called it approvingly, "a remarkable sea-dish—an intellectual chowder of romance, philosophy, natural history, fine writing, good feeling, bad sayings."[19] Such grand ambition and inclusivity mark Melville literary narrative and global cartography. Melville is not interested in mapping places, exotic or local, or in mapping the expansive space of the nation. He is interested in representing the sprawlingly vast, seemingly unrepresentable space of the world system itself.

A World Text

The degree to which *Moby-Dick* is not a traditional novel may also be viewed in its distance from the nation-state, a spatiopolitical form closely associated with the novel in European literature. Rather, as Moretti has argued, Melville's book is really a *world text.* Moretti coins this term to help explain a noteworthy characteristic of the "modern epic," a term used to name a transnational genre or subgenre which includes *Moby-Dick,* as well as Goethe's *Faust,* Wagner's *Ring,* Joyce's *Ulysses,* Pound's *The Cantos,* and García Márquez's *One Hundred Years of Solitude,* among several others. Moretti argues that these works do not fit in other genres, both because of their epic scope which extends well beyond that of the traditional novel, and because of their monumental status. They are "sacred texts" according to Moretti, monuments both to and in the linguistic and national cultures in which they were produced. Yet, for all their status as national monuments, Moretti insists that "they are all *world* texts, whose geographical frame of reference is no longer the nation-state, but a broader entity—a continent, or the world-system as a whole."[20]

Modern epic is the term Moretti prefers as the name of his super-genre, although he must emphasize each word on its own in order to make his case. He concedes, for instance, that *encyclopedia* might be the more apt term for them, following Northrop Frye's usage in *The Anatomy of Criticism,* but he preferred the "narrative connotations" of the term *epic.*[21] In everyday parlance, these works certainly are epic, in the sense of being large, covering a vast ground, and being technically dense and detailed. In some cases (most obviously in *Ulysses*), they are clearly performing functions somewhat akin to those of the ancient epic, attempting to register an entire mythic world. But they are also altogether modern, being both *of* and *for* the modern world, and as Moretti will argue, being the symbolic form of the modern world system á la Immanuel Wallerstein and Ferdinand Braudel. Their modernity may been registered in both the subject matter and the fragmentary aspects of the works. They are "flawed" masterpieces, attempting the epic totality but inevitably offering fragmented or incomplete pictures. The Homeric epic is not possible in "a world abandoned by God" (in Lukács's phrase) or a world in which "all that is solid melts into air" (in Marx's). As Melville concedes, in "Cetology," "This whole book is but a draught—nay, but the draught of a draught. Oh, Time, Strength, Cash, and Patience!"[22] Modern epics, as such, are largely unread; like Scripture, they are kept as sacred throughout the culture, including the popular culture—who in United States, if not the whole English-speaking world, has not heard of *Moby-Dick*?—but only the educated elite, largely in the scholastic institutions, actually reads them. Flawed masterpieces, sacred texts, compendia of an entire culture, modern epics are world texts, figuring forth an image of the world system.

Moretti specifically opposes the *modern epic*, a genre of *world texts*, to the novel, which is the supremely modern literary form and which is inextricably tied, by the nineteenth century at least, to the nation-state. In Moretti's view, the modern epic arises, in part, as a response to the emergence of the modern world system, which is perhaps why the modern epic also arises in areas that Wallerstein labeled the "semiperiphery." Moretti notes that Goethe's and Wagner's Germany was still divided, Melville's "America" was still in formation and without a strong sense of national identity, Joyce's Ireland was a colony, and García Márquez's Columbia was entrenched in a newly modernizing Third World. Whereas England and France, with strong national cultures and extraterritorial empires, clearly represent the "core" and establish national literary traditions of the novel, these other countries develop different forms. "The construction of national identity—henceforth required of the novel—is thus replaced, for the epic, by a far larger geographical ambition: a global ambition, of which *Faust* is the unchallenged archetype. The take-off of the world-system has occurred—and a *symbolic form* has also been found for this new reality."[23]

For Moretti, then, the modern epic is the literary form of the world system itself. But these works are not world texts merely because they have a multinational continent or the world system as their "geographical frame of reference," but also, as I suggested above, because they attempt to include an entire world. Moretti notes that the modern epic employs *bricolage*, Claude Levi-Strauss's term for the cultural work of putting things together that "lay to hand" rather than the rational engineering of a thoroughly planned out project. This aspect of modern epics is visible in their intermingling of high and low, of classical antiquity and dime-store pop culture (see how it is in Eliot's *The Waste Land*). It is also visible in the tremendous historical and geographical range of these works, whose representations may bounce between the all-too-present and the immemorial past, between a modern New York or London and a distant Polynesia or Egypt. The modern epic is encyclopedic in scope and range, but employs collage as its artistic technique. In these works the polyphony and heteroglossia that Mikhail Bakhtin had located in the novel frequently explode in a cacophony of voices—*Faust*'s *Walpurgisnacht* scene and *Moby-Dick*'s "Midnight—Forecastle" are the supreme examples—and, with the adoption of the stream of consciousness technique, even a solitary narrative voice may in fact be polyphonous (in *Ulysses* or *One Hundred Years of Solitude*). In summation, the modern epic is the symbolic form of the world system, in large part, because it is the form that attempts to encompass the entire world, its history, geography, artistic forms, and on and on.

For Moretti, all this indicates the profound ideology of Eurocentrism, or more precisely, of the West's domination of the rest of the world, and thus the modern epics in question become the symbolic embodiment of modern imperialism. I do not agree on this point, except insofar as a modern imperialism is part of what is being represented in the world texts, and to fail to include it in the overall representation would be to leave a crucial element of the world system unmapped. In the next chapter I give a specific example of how, in my estimation, Moretti fails to read the more salutary effects of *Moby-Dick*'s use of polyphony, but I can say here that Moretti underestimates the degree to which these "sacred texts" may resist the very idolaters who have placed them prominently at the national altar.

In my use of the term *baroque*, I mean something similar to what Moretti identifies in his modern epics. (Indeed, I believe that the baroque age's *Don Quixote* could be a precursor to *Faust* and the others, although Moretti would probably disagree.)[24] As I discussed in Chapter 1, the baroque is associated with the birth of the modern, along with the birth of capitalism, the modern novel, the modern state, modern imperialism and urbanism. But it also has a distinct though related meaning involving complexity, difficulty, allegory, and extravagance but also involving the high art and philosophy (at least, once the term was rehabilitated by modernists like Walter Benjamin). All of these labels, for good or for ill, can also be attached the social, political, artistic, and literary forms the emerged in the baroque epoch. Thus, in the end, the term *modern epic* falters slightly, as it seems to ossify these works in their sacred shrines, and does not fully explore the degree to which their force may be used to understand our own world. The rise and dominance of the nation-state is not the end of the story, and the emergence of the postnational world system, which Melville and others have attempted to map, now forms the basis for literary as well as political representation. "It is indeed the new world system [. . .] which is for us the absent totality, Spinoza's God or Nature, the ultimate (indeed, perhaps the only) referent, the true ground of Being of our own time. Only by way of its fitful contemplation can its future, and our own, be somehow disclosed."[25]

In Melville's case in particular, this contemplation manifests itself in a baroque and encyclopedic work that attempts to supersede the national in aesthetically representing a postnational world in which "the meanest mariners, and renegades, and castaways" are ascribed "high qualities" and are woven all around with "tragic graces" precisely because the tragedies of this world system know no national borders. The world system that Melville sees in its emergence, the same one Marx registers in his analysis of the world market, is the system of globalization in which we now live and which we attempt, in one way or another, to make sense of, or give form to: in other words, to map. To the extent that national literatures—including the national, local, personal, literary, and other narrative forms within them—function to produce maps of the world in which the national is prominent, and in which the other areas of the maps are figured precisely in relation to the nation-state, in the era of globalization, we can heartily agree with Goethe, and proclaim: "National literature is no longer of any importance; it is time for world literature, and all must aid in bringing it about." In *Moby-Dick*'s baroque art, a single world text offers us a fine example of that postnational world literature best suited to our time.

Chapter 4

Anti-Ishmael

"Call me Ishmael." This invocation to the reader of *Moby-Dick* is, perhaps, the most famous line in American literature, and part of its notoriety, no doubt, lies in its being a famous *first* line. Even those who have never read *Moby-Dick* nevertheless *know* that it begins with the words "Call me Ishmael." Of course, those who have read the novel also know that the three-word sentence appears as the opening line of "Loomings," Chapter 1 of *Moby-Dick*, but not necessarily as the beginning of the novel. Preceding "Loomings" are two unnumbered chapters, "Etymology" and "Extracts,"[1] chapters rarely discussed in most critical treatments of the novel. When they are discussed, it is usually *en passant*, a brief mention before getting to the meat of the interpretation, which more often than not revolves around the figure of Ishmael. By reading "Call me Ishmael" *as* the beginning of *Moby-Dick*, a long line of literary critics has established that Ishmael must be the central figure of the novel. Of course, Ahab will become the tragic hero or villain of *Moby-Dick*, but Ishmael will be the "voice" of the novel.

This Ishmael-centric tradition within Melville studies transcends different movements and ideological positions. From the Melville revival of the 1920s, through the period of academic and cultural canonization in the 1940s and 1950s, and persisting with the so-called New Americanists of recent years, the "Call me Ishmael" point of departure and the line of thought that extends from it have remained the prevailing perspectives. Meanwhile, the unnumbered chapters that precede "Loomings" have been almost entirely ignored.[2] In this chapter, I propose that the Ishmael-centric line of thought has enabled an interpretation of *Moby-Dick* that supports a nationalist cultural program, transforming the novel into a national narrative and ignoring its *postnational* force.[3] By reading "Etymology" and "Extracts" as the novel's true point of departure, I disclose a different *Moby-Dick*, one which resists encapsulation in a national narrative and anticipates the era of globalization or of the world system of which we are now a part. This *Moby-Dick* put forward an overwhelming figure of the whale, rather than a representative national figure, and through its iconography, the novel overwhelms the very national narrative to which it has be put to use in American Studies. I believe that a postnational *Moby-Dick*, and by extension a postnational approach to American Studies, proceeds from this alternative line of thought in the text. This chapter will provide a reading of *Moby-Dick*'s postnational beginnings, and bring to light the nomad thought of the novel, explored in greater detail in Chapter 5.

Ishmael as American Adam

Beginning with Ishmael, or establishing Ishmael's centrality at the beginning, enables a reading of *Moby-Dick* as an essentially American national narrative, a narrative which is designed to reproduce an ideological *mythos* of a special, even divinely ordained, status in the world, such that the nation may be considered unique, distinct from the rest of the world and, in turn, *better* (e.g. more free, more moral) than other nations. As discussed in Chapter 2 above, national narrative both presupposes and helps to create an "imagined community," in Benedict Anderson's memorable phrase, a collective subject called a nation. The Ishmael-centric reading so prevalent in Melville Studies establishes a national narrative for *Moby-Dick* in part by focusing on the individual, a figure who can become representative of the nation, much as Natty Bumppo served that purpose for Cooper. In viewing *Moby-Dick* as a kind of personal narrative, even though the formal convention does not hold throughout the novel, Ishmael-centric critics have helped to establish the novel as a national narrative by allowing the individual to stand for the nation. As Wai Chee Dimock argues, a "poetics of individualism" is central to the ideological regime of nation-formation and empire-building in the United States.[4] That is, the sovereign individual is representative of the nation, which is geographically and ideologically extended over distant domains in what Dimock, following Thomas Jefferson, calls an "empire for liberty." American national narrative celebrates this movement from the individual to the national and on to the imperial, a movement that is unified in purpose and effects. Thus, a seemingly personal narrative by "one Ishmael" can easily become a national narrative.

This movement is visible in the familiar images associated with American national narrative. "The image repertoire productive of the U.S. national community can be ascertained through the recitation of its key terms in the national meta-narrative commonly understood to be descriptive of that community. Those images interconnect an exceptional national subject (American Adam) with a representative national scene (Virgin Land) and an exemplary national motive (errand into the wilderness)."[5] An Ishmael-centric reading of *Moby-Dick* fits neatly into this national narrative framework. Notwithstanding his evocative name, Ishmael conforms to the principle of the Adamic American hero, innocently voyaging into experience. Many Ishmael-oriented critics have seen *Moby-Dick* as, at least in part, the story of a nation-representing Ishmael triumphing over a nation-destroying Ahab.[6] Much of twentieth-century criticism has hailed, or more recently assailed, *Moby-Dick* for being a national narrative *par excellence,* and such a view, I believe, is firmly grounded in seeing Ishmael as an American Adam.

However, in an odd sense, a largely unsympathetic audience of mid-nineteenth-century critics were closer to the mark in identifying the power, and the problems, of the baroque *Moby-Dick.* A number of contemporary reviewers saw *Moby-Dick* as a failed personal narrative, chaotically multiformal and exceeding the boundaries of both its genre and good taste. Although Melville's friend Evert Duyckinck relished the multiple forms, declaring the novel "a remarkable sea-dish—an intellectual chowder of romance, philosophy, natural history, fine writing, good feeling, bad sayings," even he finds Ishmael inconsistent, exhibiting "painful contradictions," with "a mind driven hither and thither like a flame in a whirlwind," who cannot in the end secure "much admiration" from the reader.[7] A less charitable reviewer noted that *Moby-Dick* is "distressingly marred by an extravagant treatment of the subject" and that the "style is maniacal—mad as a March hare—mowing, gibbering, screaming, like an incurable Bedlamite, reckless of keeper or strait-waistcoat."[8] Another claimed that Melville was "gauging, at once, our gullibility and patience," and that Melville's writings since *Typee* had "shown a decided retrogression" with respect to "all those qualities that make books readable."[9] As I discussed in Chapter 1, the view of *Moby-Dick* as a flawed work (albeit, perhaps, a flawed masterpiece), comes much closer to seeing the real power of Melville's baroque excess than the fawning idolatry of Melville's twentieth-century cheerleaders. *Moby-Dick* does, in fact, present a kind of failed personal narrative as part as its critique of the nationalist project dominating letters and political discourse in the mid-nineteenth-century United States. The global, literary cartographic project of *Moby-Dick* is not the story of a "Whaling Voyage by one Ishmael," as suggested in "Loomings." By recognizing the degree to which the novel was "an ill-compounded mixture of romance and matter-of-fact" and "an odd book, professing to be a novel, wantonly eccentric,"[10] many of Melville's contemporaries, in their very displeasure and outrage, validated the proposition that *Moby-Dick* is not representative of the national culture or national narrative form, but that it wanders out of bounds, becoming a savage anomaly that perhaps undermines the national ideal.

As William Spanos's "destructive" reading of the novel's "errant art" demonstrates, *Moby-Dick* contains the weapons to combat the logic of this national narrative of a Puritan errand into the wilderness. Spanos, however, does not displace Ishmael from his central position as narrator and observer. Rather he uses Ishmael to counter the power arrayed on behalf of *Moby-Dick* as national narrative, and in so doing, he discovers an Ishmael quite different from the representative American national subject. Spanos points out that the "Call me Ishmael" invocation cannot be taken to mean the same thing as "My name is Ishmael." The indeterminacy of the former phrase orients "the reader/listener's attention to the possibility that the name is a mask [. . .] By calling attention to the fictional status of the narrator, to his name *as a sign*, Melville simultaneously reiterates and breaks down the distinction between the privileged author and the first-person narrator he 'creates' to tell his story."[11] In Spanos's reading, then, Ishmael is not the subject who stands above or outside of the narrative; nor is he the representative of a fixed identity (i.e. the American Adam). If Ishmael's narrative is *errant*—erratic and extravagant, not to be reduced to a ruling *logos*—at least his is no errand into the wilderness. Spanos discovers in Ishmael's errancy a counter-hegemonic force opposed, not specifically to Ahab's "tremendous centralization," but to the "ontotheological tradition" that undergirds the organizing *mythos* of the American jeremiad, whose figure in *Moby-Dick* appears in Father Mapple's sermon.[12] Ishmael vs. Mapple, not Ishmael vs. Ahab: this is the antagonism which, in Spanos's view, energizes *Moby-Dick*, and which draws the lines of battle between a posthumanist and postmodern errancy and the Enlightenment theo-logocentrism *cum* anthropo-logocentrism animating national narrative.[13]

Among Spanos's most emphatic points is that the Ishmael narrative is fundamentally errant, irreducible to a totalizing or imperial narrative structure. Spanos usefully retrieves the Ishmael figure from the hands of both the Cold War critics, for whom Ishmael stands as the representative American Adam fighting Ahab's totalitarianism, and the New Americanists, for whom Ishmael is complicit in the American nationalist or imperialist mission. But I believe that Spanos relies too heavily on Ishmael for his argument. That is, the errancy of *Moby-Dick* need not be, and I think *is not*, specifically Ishmael's. Moreover, to focus on Ishmael proves misleading. Although Spanos believes that the "Ishmael" name is just a *sign*, a name "as good as any" to initiate the story,[14] this sign eventually oversignifies, becoming (in the minds of more than a few critics) its own totalizing narrative voice.

One of the chief problems with the Ishmael-centered interpretation of *Moby-Dick* is that, by assuming a fixed identity and a stable narrative voice in the figure of "Ishmael," critics tend to deny the polyphony and heteroglossia (or the very *errancy* Spanos celebrates) that characterizes so much of the text. Even Franco Moretti, who acknowledges and praises the polyphony of *Moby-Dick*, at its most dramatically audible in "Midnight—Forecastle," winds up situating Ishmael as the *one* voice of narration. Moretti argues that *Moby-Dick* repeatedly reduces the polyphony to silence and that, by silencing the multiple voices heard especially in earlier parts of the novel, *Moby-Dick* is complicit in endorsing the "leveling" processes of imperialism and capitalism.

> Here, the monologic device is the voice of Ishmael (Ishmael as narrator): *one* voice, omnipresent and situated at a level where no other can respond to it, and thus call it into question. In the central part of the novel, roughly half

> of the chapters are in fact concluded by the narrator's reflections—a strategic placement, investing Ishmael's words with a particular authority [. . .] In the last resort, the ambition of the narrator of *Moby-Dick* is precisely this: to take the multifarious codes of nature and culture, and to demonstrate that they are all to be found in the moral super-code. To take polyphony, in other words, and reduce it to a single language: ultimately, to eliminate it altogether.[15]

Obviously, Spanos would dispute this interpretation, arguing instead that these chapters reveal Ishmael's "carnivalesque parody" of such a super-code.[16] However, by taking for granted the centrality of an "Ishmael" as the narrator, critics like Moretti assume that this singular and authoritative narrative voice speaks throughout *Moby-Dick*, that the various voices heard in the novel are subsumed or silenced by the voice of one Ishmael. As I will argue, the Ishmael-centric approach to the novel leads, or misleads, inevitably to a national narrative view that necessarily disregards the powerful postnational energies of *Moby-Dick*.

Philological Beginnings

This reading, or misreading, follows directly from taking "Call me Ishmael" as the beginning of *Moby-Dick*. The *beginning* is not merely the place one starts reading; it is the place that conditions the reading as a whole. As Edward Said has noted, "when we point to the beginning of a novel, for example, we mean that from *that* beginning in principle follows *this* novel."[17] In other words, the inevitable question of where to begin is also and by necessity a question of meaning, since the meaning of a given text will be determined, in large part, by the beginning. The beginning is an *Ansatzpunkt*, as Erich Auerbach has called it, a point of departure from which the entire voyage unfolds. From the writer's perspective, choosing to begin at a certain point is also choosing to produce a certain work. From the reader's perspective, emphasizing one point of departure over another conditions how the entire work will be read. With "Call me Ishmael" as the beginning, the interpretation of *Moby-Dick* that follows must assert that the story is Ishmael's, that the narrator is singular, and that the narrative's authority lies in this character. Anything that does not seem to fit within the Ishmael-centric plot is then perceived as a digression. Yet if "Call me Ishmael" is not the *Ansatzpunkt* of *Moby-Dick*, then a different story altogether may emerge.

In "Philology and *Weltliteratur*," Auerbach notes that "The characteristic of a good point of departure [*die Ansatz*] is its concreteness and precision on the one hand and, on the other, its potential for centrifugal radiation."[18] Said elaborates Auerbach's argument by noting that a point of departure has two distinct aspects that animate one another: "One leads to the project being realized: this is the transitive aspect of the beginning—that is, beginning with (or for) an anticipated end, or at least an expected continuity. The other aspect retains for the beginning its identity as *radical* starting point: the intransitive and conceptual aspect, that which has no object but its own constant clarification."[19] As a point of *departure*, the *Ansatzpunkt* always indicates a movement, a trajectory, a tendency toward a goal; hence, its potential for centrifugal radiation. As a *point* of departure, it must also concentrate on itself, returning to its place of origin in an intense meditation on the concrete, precise, discrete identity, never losing sight of itself as its momentum unfolds. As Said explains, "The transitive mode is always hungering, like Lovelace perpetually chasing Clarissa, for an object it can never fully catch up with in either space or time. The intransitive, like Clarissa herself, can never have enough of itself—in short, expansion and concentration, or words in language, and the Word."[20]

The *Ansatzpunkt* for *Moby-Dick,* it seems to me, cannot be Ishmael, but is in fact directly legible in its original or alternative title: *The Whale.* Beginning with "Etymology" and "Extracts," we see the beginning in "the whale" itself, which is defined, presented in various languages, and elaborated through centuries of discourse across continents. Just as in Said's characterization of points of departure, the novel will ceaselessly pursue the whale—a particular one whom we do not see in the flesh until Chapter 133, but also the entire species about which we hungrily gobble up information and knowledge—but in the novel we also forever return to the whale, reiterating its manifold meanings and symbols. That is, *Moby-Dick* will explore both "words in language" (whales) and "the Word" (Whale).

The Whale maintains a seemingly deliberate ambiguity throughout the novel, owing partly to the shifting nuances of the article *the* that allow it to be both general and specific at once. Thus, for example, "the *whale*" of "Cetology" describes the general character of the animal, which can be subcategorized into its various species; "*the* whale" of the chapter "Moby Dick" refers to a specific character in the novel, complete with a personality and a legendary biography. This ambiguity brings together the narrative and the descriptive or scientific elements of *Moby-Dick* under one sign. The reader soon learns that ambiguity or indeterminacy or polysemy or "inscrutability" (as Ahab would have it) is a crucial aspect of *the whale* and of the novel. Already in "Etymology" and "Extracts," the concrete *Ansatzpunkt* is revealed and the centrifugal movement begins. Before we have met an unknown but still reasoning thing that bids us call him Ishmael, before we are given a monologic device to clarify or impose meaning on things set before our eyes, the unnumbered chapters announce that *the whale* is the beginning of *Moby-Dick.*

With its own "Usher" (to whom I'll return below), "Etymology" ushers *the whale* into being in the context of language. This beginning sets the tone for the novel as a whole. Throughout the novel, Melville associates the whale with language, perhaps most memorably in "Cetology," with its whale-as-book taxonomic scheme (e.g. *Folio* whales being the largest, then *Octavo, Duodecimo,* and so on.). The beginning of *Moby-Dick,* in fact, involves a philological exploration of the whale. The world of the whale, as Melville presents it, is contained in the fragile density of the Word (in "Etymology") and born in the material flesh of discourse (in "Extracts"). From this point on, the novel manifests the transitive and intransitive effects that characterize every beginning: a headlong rush towards *the* whale and a repeated meditation on the *whale.*

The etymological explication of the whale starts with a reference to Richard Hakluyt (spelled "*Hackluyt*").[21] The line attributed to Hakluyt makes specific reference to education, language, and truth—in short, to knowledge:

> While you take in hand to school others, and to teach them by what name a whale-fish is to be called in our language, leaving out, through ignorance, the letter H, which almost alone maketh up the signification of the word, you deliver that which is not true.

Thus the etymological beginning of *Moby-Dick* opens with a warning, a caveat that will be repeated throughout the book: knowledge of the whale, to be learned and to be disseminated, is (or is in perpetual danger of being) incomplete and false. Moreover, and this is crucial in understand the global extent of Melville's baroque project, by leaving something out, one has inevitably delivered something that is not true. Small wonder then that the novel, which some see as the story of an Ishmael or an Ahab, turns out to be a work of such "outreaching comprehensiveness of sweep, as if to include the whole circle of the sciences, and all the generations of whales, and men, and mastodons, past, present, and to come, with all the revolving panoramas of empire on earth, and throughout the whole universe, not excluding its suburbs."[22] With the Hackluyt warning right on the very first page, Melville announces that the novel is no mere personal narrative of adventure. *Moby-Dick,* whose *Ansatzpunkt* is the whale itself, is really about the entire world.

This opening caveat puts one in mind of others scattered throughout the novel, including those where Melville warns of the direct, physical dangers of acquiring accurate knowledge of the whale. In "Of the Monstrous Pictures of Whales," for instance, the narrator cautions that "the only mode in which you can derive even a tolerable idea of his living contour, is by going a whaling yourself; but by so doing, you run no small risk of being eternally stove and sunk by him Wherefore, it seems to me you had best not be too fastidious in your curiosity touching this Leviathan." This ironic attitude towards scientific or artistic knowledge is maintained throughout *Moby-Dick.* The reader must withstand a barrage of facts, data, descriptions, and documentary details, such that the novel appears, at times, more like a loosely gathered, nonfictional study of whales and whaling, as the critics so frequently complained; yet the reader is also confronted, again and again, with admonitions or warnings against believing that they have got some clear grasp of the subject. Indeed, the novel repeatedly tells us that, notwithstanding the quasi-scientific descriptions abounding in the text, complete knowledge of the whale is impossible. At a notorious moment in "Cetology," for instance, the narrator concedes that he is "the architect, not the builder" of the system of knowledge of the whale; the "cetological System" must remain unfinished, incomplete, "the draught of a draught"; in "The Tail," the narrator admits, "Dissect him how I may, then, I but go skin deep; I know him not, and never will."[23] If, "through ignorance," something is left out, the overall rendering of the whale will not be true, at least to the extent that the truth is ever arrived at in *Moby-Dick*. The invocation of Hakluyt announces a project that involves the writer and the reader in a study of *the* whale and of *whales*, of which Moby Dick is a unique specimen. But it also declares that the project must remain unfinished. Like the world system it assiduously tries to map, the baroque project of *Moby-Dick* remains a project, an ongoing effort in representing the unrepresentable.[24]

Because it is the object toward which the narrative rushes and because the species is marked and remarked upon continually, the whale is forever resignified. If the letter *H* is left out of the word *whale*, the word becomes *wale*, which is a mark on the body or skin made by, especially, the stroke of a whip, the visible trace of a body having been struck. Through Hakluyt, then, a sign is introduced that, metonymically, anticipates the figure of Ahab, the figure of *the* whale, and the figure of *wales* in general.

> [Ahab] looked like a man cut away from the stake, when the fire has overrunningly wasted all the limbs without consuming them [. . .] Threading its way out from among his grey hairs, and continuing right down one side of his tawny face and neck, till it disappeared in his clothing, you saw a slender rod-like mark, lividly whitish. It resembled that perpendicular seam sometimes made in the straight, lofty trunk of a great tree, when the upper lightning tearingly darts down it, and without wrenching a single twig, peels and grooves out the bark from top to bottom, ere running off into the soil, leaving the tree still greenly alive, but branded. Whether this mark was born with him, or whether it was the scar left by some desperate wound, no one could certainly say.[25]

The "marked man," Ahab, is brought into relation with the "whale-fish" on the first page of the novel.[26] Moby Dick, too, is marked, in many senses of the word. But Ahab, distinguished by his *wale*, may be said to be lacking the heart of the whale, the heart of the word "which almost alone maketh up the signification of the" *whale*.

Consider also the way in which this marking and remarking anticipates the appearances of the whale—both *the* whale and the *whale*—later in the novel.

In "The Whiteness of the Whale," the narrator tries to come to terms with *the* whale, Moby Dick, through an examination of its most visible distinguishing mark, its whiteness. This attribute is explored through literary, mythological, historical, anecdotal, and scientific references, all leading to the conclusion that both *the* whale and the *whale* are inscrutable and overdetermined. Though he does not know "where lie the nameless things of which the mystic sign gives forth such hints," the narrator nevertheless can assert that "of all these things the Albino whale was the symbol." These multiple meanings inhere in the whale itself. Elsewhere, in "The Blanket," the narrator *reads* the whale through the "numberless straight marks in thick array [. . .] engraved upon the body itself." These marks are "hieroglyphical," imbued with meanings that the reader cannot quite understand: "the mystic-marked whale remains undecipherable."[27] The marks, or *wales*, of whales render them full of meaning and yet inscrutable.

A *wale* also refers to any one of a number of strong strakes or wooden bands on the hull of a ship used to strengthen or fortify the structure of the vessel (e.g. as in the term *gunwale*). In a sense, then, the Hakluyt quotation may be said to prefigure the multiple confrontations—leading, of course, to the ultimate confrontation—between the whale and the ship in *Moby-Dick*: the force of *the whale* versus the strength of the *wales*. "Retribution, swift vengeance, eternal malice were in his whole aspect, and spite of all that mortal man could do, the solid white buttress of his forehead smote the ship's starboard bow, till men and timbers reeled."[28] Moby Dick, who had once "dismasted" Ahab, now dismasts and destroys the entire ship, cracking its timbers and wales. The *Pequod*, like Ahab, will lack that heart of the whale most essential for delivering that which is true. In the opening Hakluyt quotation, the *Pequod* is itself metonymically or synecdochically invoked by reference to its wales; it is at once connected to and distinguished from the whale. In one brief comment, presented as part of an etymology of the word that will serve as *Ansatzpunkt* for the entire work, the whale, the captain, and the ship are figured forth.

Long before the white whale that marked and is marked by Ahab rams into the *Pequod*, "Etymology" invokes definitions that already flesh out the whale and the conflict. First, *Webster's Dictionary*: "WHALE. * * * Sw. and Dan. *hval*. This animal is named from roundness or rolling; for in Dan. *hvalt* is arched or vaulted"; *Richardson's Dictionary* then adds, after attributing "whale" to Dutch, German, and Anglo-Saxon sources, "to roll, to wallow." The etymological origins of the whale, *roundness* and *rolling*, radiate throughout the novel, as descriptions not only of whales in *Moby-Dick*, but also of the world of *Moby-Dick*. The most prominent example, perhaps, may be seen in the final conflict between the white whale and the *Pequod*, inasmuch as the "roundness" and "rolling" come to mark the features of the novel's end: "And now, concentric circles seized the lone boat itself, and all of its crew, and each floating oar, and every lance-pole, and spinning, animate and inanimate, *all round and round* in one vortex, carried the smallest chip of the Pequod out of sight [. . .] Now small fowls flew screaming over the yet yawning gulf; a sullen white surf beat against its steep sides; then all collapsed, and the great shroud of the sea *rolled on as it rolled* five thousand years ago."[29] The definition of the whale, its literal and etymological origins, are repeated, but with very different effects, in the final moment of the book. The overwhelming nature of whale, an unrepresentable totality, literally overwhelms the narrative of Ahab, the *Pequod*, and, of course, an Ishmael.

"Etymology" establishes this roundness with linguistic circularity. It marks a point of departure for a line of thought leading to a conclusion quite different from the one in which Ishmael bobs up to the surface from a whirlpool that consumes everyone else, a "survivor's narrative" that extends from "Call me Ishmael" to the Epilogue.[30] With "Etymology," the novel sets out on a rounded, circular or vorticist-helical[31] trajectory in which the narrator "reads" *the* whale in terms of the *whale* (and vice-versa) while reading the *wales* that serve as marks or remarks. Tracing the line from "Etymology" to the last lines of "The Chase—Third Day," the very figure of the whale seems to undermine, or literally *overwhelm* or *roll-over* the personal and national narratives that appear to begin with the opening sentence of "Loomings." By starting with "Etymology," we see a *Moby-Dick* that is not the tale of a young man's errand into the wilderness and his providential return to safety, but the roiling epic of a world of error, of uncertainty, and of horror. The story of "one Ishmael" cannot as such achieve the goal of a global, round and rolling, representation. Rather, the evocative image of Melville's epic is the supra-individual whale as world itself.

The World in a Word

The languages in which "Etymology" presents the word *whale* reflect the tremendous breadth of this world and the scope of the novel: thirteen languages, from ancient to modern, from Middle Eastern to European to South Pacific, frame an image geographically and historically.[32] Thus, "Etymology" begins to trace a line of thought quite distinct from the narrative trajectory of an Ishmael-centered plot. For the very multiplicity of languages, like the multiple "renderings" of the text of the doubloon later, suggests that no one language—and, by extension, no one speaker—is alone authoritative.[33] The indeterminacy of a "true" name, a "true" language, or a "true" voice prefigures the indeterminacy of the *whale*, as well as of *the* whale and the *narrator*. The interpretations of *Moby-Dick* which posit Ishmael as the narrator, and which thereby invest in Ishmael a singular authority, must ignore (and certainly *have* ignored) this profound indeterminacy. But "Etymology" makes visible, by its very excesses, the bankruptcy of a reading of *Moby-Dick* that requires a spokesperson able to speak a universal, representative language of fixed and stable meanings. By undermining the role of the American Adam—the ideological servant of national narrative, who would represent an "imagined community" of readers or listeners—"Etymology" substitutes an image of a multinational world system for the nationalist image, a global Babel.[34] With its *whale* in thirteen languages, ordered chronologically (at least from a certain Western point of view) and ranging widely over the globe (from the Mediterranean to the North Atlantic and Scandinavia, and on to the South Pacific), "Etymology" attempts to capture the entire world of nations: the world in a word.

In "Extracts," the philological project expands to a general discourse of the *whale*, affording "a glancing bird's eye view of what has been promiscuously said, thought, fancied, and sung of Leviathan, by many nations and generations, including our own." If "Etymology" ushers the whale into being, discovering the outline of a world within a word, "Extracts" gives birth to the whale in the material flesh of discourse. As in "Etymology," the ordering is basically chronological, such that the extracts promulgate the notion of a universal history, from ancient to modern. The geographical movement also appears to be largely the same: from the Orient (the Biblical Holy Land, corresponding to the Hebrew and Greek of "Etymology") to the west and north (from the Mediterranean to northern Europe) and farther afield until one finds oneself singing with the whalemen of the South Pacific fishery. In addition to its historical and geographical scope, "Extracts" also introduces the Bakhtinian, multiformal character of *Moby-Dick*, inasmuch as it represents all manner of forms or "stylistic unities," literary and non-literary, poetic and scientific, travelogue, oratory, law, and so on. As Bakhtin notes, "the stylistic uniqueness of the novel as a genre consists precisely in the combination of these subordinated, yet still relatively autonomous, unities (even at times comprised of different languages) into the higher unity of the work as a whole: the style of the novel is to be found in the combination of styles; the language of the novel is the system of its 'languages.' "[35] The various forms included among the extracts—including the sublime (as in *The Faerie Queene* excerpt) and the mundane or commonplace (as in Cuvier's "The whale is a mammiferous animal without hind feet")—proliferate throughout the novel, forming, as the "higher unity of the work," a sprawling image of the world, including, of course, the literary world of these forms.[36]

The collection of extracts says a good deal about whales. There is one exception, however, among the extracts—a quotation that deals with nothing cetological at all. "By art is created that great Leviathan, called a Commonwealth or State—(in Latin, Civitas) which is but an artificial man."[37] While it is certainly not surprising to see the most famous use of the word *leviathan* since the Book of Job included among the extracts (indeed, its absence would have been a shock), the reference to the state, rather than to the whale, is striking here. Melville, or that "poor devil of a Sub-Sub" who has putatively assembled these fragments, has buried this line from Hobbes's baroque treatise on sovereignty amid a vast textual history of the whale, extending from Genesis to the mid-nineteenth century. "Extracts," like "Etymology," presents a multinational world system, expansive both geographically and historically. The nation-state, that embodiment and aim of national narrative, is subsumed within this world system.[38] Melville's Leviathan is thus not Hobbes's. Hobbes is concerned with the nature and functioning of sovereignty, and *Moby-Dick* disrupts and calls into question the ideas of sovereignty and, in particular, the sovereign subject, as I will discuss in the next chapter. The extensive multinational frame, which is constructed in the first two, unnumbered chapters, ranges across both space and time, and displaces the supposed centrality of the sovereign subject and of the nationalist project. In other words, *Moby-Dick*, through its very excess, through its "outreaching comprehensiveness of sweep," announces in its opening pages that it is not, and cannot be, the representative American national narrative that it has become for many; rather, it is a baroque cartography of a world system within which the American imperium and American national narrative, with its representative national subject, are overwhelmed and subsumed. The philological beginnings of *Moby-Dick* invite us to imagine the novel in the multinational context of a baroque complex of language and geography. The novel's forces lies in its attempt to grapple with the representational crisis occasioned by a true leviathan, the emerging, postnational world system.

Call *Who* Ishmael? A Note on Conceptual Personae

By taking "Etymology" and "Extracts" as the beginning of *Moby-Dick*, I have been arguing for a line of thought in the novel that runs counter to the Ishmael-centered line which extends from the first three words of "Loomings" to the Epilogue and which has served as the basis for a reading of the novel that turns it into a national narrative. As I suggest above, many critics have imposed a peculiar coherence and narrative authority on the text by insisting that a monologic Ishmael is the narrator. By ignoring the unnumbered chapters, these critics have also overlooked the frequent unraveling of the narrative acts, the subversion of the narrator's authority, and the ambiguities of the text in favor of an overarching master code (Ishmael's "moral super-code," as Moretti calls it). As Spanos's reading makes clear, however, Ishmael need not be viewed as that monologic authority. Indeed, Ishmael need not even be considered a singular presence in *Moby-Dick*. "Ishmael" might be thought of as one of many *conceptual personae*, to borrow a term from Deleuze and Guattari.[39] His persona is not that of a "character" in the traditional sense; rather, he is a figure that accompanies concepts, a figure through whom thought moves, like Nietzsche's Zarathustra or Dionysus, for example. Ishmael is not the only such figure in *Moby-Dick*. Before anyone has us call him "Ishmael" in "Loomings,"

"Etymology" introduces us to the "pale Usher," whose etymology ushers the whale into being on the novel's first page; then "Extracts" introduces us to the "poor devil of a Sub-Sub" librarian, whose extracts, culled from "the long Vaticans and street-stalls of the earth," locate the whale in an expansive history and geography of discourse.[40]

Already placed in brackets—"the typography that disguises," as Claude Richard[41] puts it—neither of these figures makes a good candidate for the American Adam so pivotal in American national narrative. Each sojourns through a vast terrain of oral and written language, hardly "Virgin Land" waiting to be discovered and conquered. Their forays into the worlds of dictionaries and libraries certainly do not resemble an "errand into the wilderness" in any sense that would link either of them to a Providential mission. Rather, they tend to block the gateway to national narrative, and open the text to a polyphonic and polysemic world system beyond the image repertoire of Americanism.

> [The pale Usher—threadbare in coat, heart, body, and brain; I see him now. He was ever dusting his old lexicons and grammars, with a queer handkerchief, mockingly embellished with all the gay flags of all the known nations of the world. He loved to dust his old grammars; it somehow mildly reminded him of his own mortality.]

The "I" in this passage does not enjoin the reader to call him (or her) Ishmael; the narrator is as yet unnamed. And the Usher is clearly not the Adamic figure of innocence and futurity; he does not elicit an optative mood in either speaker or reader. In fact, this minor custodian of knowledge, now apparently dead, appears to have had only one pleasure in life: the contemplation of his own finitude. Brushing dust off old dictionaries, those compendia of dead languages and displaced meanings, with a cloth "mockingly embellished" with a symbolic, international political geography, he does not invite celebration in all things American. He is as mocked by the variety and extensiveness of the world as he is consumed by it. *All* the flags of *all* the known nations of the world—rather than decorating individually and specifically, the flags are signs of conglomerating global immanence, an ever-present multinational perspective that consumes and subsumes Adamic figures and leviathans alike.

The Sub-Sub-Librarian is a persona expanding that of the late consumptive Usher just as "Extracts" expands on "Etymology." The Sub-Sub, himself another minor custodian of knowledge, operates in a somewhat larger arena (libraries instead of a schoolhouse), but he may be an even lesser personage (being, after all, a *sub*-sub). He is no more a figure of hope and futurity than is his predecessor. Introduced as a "mere painstaking burrower and grub-worm," whose collection of extracts cannot be trusted as accurate knowledge, as "gospel cetology," the Sub-Sub belongs to a "hopeless, sallow tribe," and—though not yet dead—his kindred spirits in mediocrity are reserving his place in the hereafter. He too is consumed, or subsumed, by the immensity of the world system, by "many nations and generations." Copier of documents, bearer of dead letters, he prefigures Bartleby more than Ishmael.

When the Ishmael mask is donned in the opening line of "Loomings," then, the reader is already familiar with conceptual personae who are neither Adamic nor national. Ishmael, furthermore, is dissociated from any determinate, and determining, identity by the "Call me . . ." formulation, as Spanos demonstrates. His evocative name and the allure of his naming is so strong that, even when Ishmael, as character *or* as narrator, disappears or seems to disappear for long stretches of time during the course of the novel (e.g. when there is narration of events or phenomena that Ishmael could not have known about, such as monologues inside the mind of Ahab or scenes behind closed doors to which Ishmael could not have had access), most readers persist in naming the narrator "Ishmael," in thinking of him as *one* voice, and in assuming, despite a wealth of evidence to the contrary, that the narrator is one and the same person throughout *Moby-Dick*. However, not only will the novel be polyphonic, but the narrative voice (or voices) will be as well.

While the Ishmael mask may not be entirely removed, *Moby-Dick* presents other personae as the situation warrants: the narrator of "The Fossil Whale" is a geologist (456); he becomes a tattooed poet in "The Bower of the Arsacides" (451); and, in relating "The Town-Ho's Story," he is suddenly transplanted to Peru, presumably speaking Spanish with cigar-smoking Spaniards[42] (243). It is possible, of course, that all of these narrative voices can be ascribed to one Ishmael. Yet even where the narration seems left in the hands, or voice, of one person, polyphony persists. The monological Ishmael, whom Moretti finds in so many of the central chapters of *Moby-Dick*, is at most a parody. The narrator's "authority" in these chapters appears again and again to be a ruse, as if Melville is poking fun at the authorities and at authority itself. For all of the assumed authority of the cetological chapters' narrator (or narrators), few readers really accept the authority at face value, and it certainly seems clear that Melville does not intend them to do so. The most "authoritative" comments often provide moments of humor. So, for example, when the narrator of "The Fountain" insists that "from the heads of all ponderous profound beings, such as Plato, Pyrrho, the Devil, Jupiter, Dante, and so on, there always goes up a certain semi-visible steam, while in the act of thinking deep thoughts," it is difficult for anyone to take this seriously. (When the narrator vouches for the theory by saying how he once saw steam rising from his own head "while composing a little treatise on Eternity," and he also admits this occurred "after six cups of hot tea in my thin shingled attic, of an August noon," any doubt as to the parodic nature of this "authority" is definitively removed.)[43] Moretti chooses to overlook the deliberate irony and therefore overlooks not only the persistence of polyphony in the cetological chapters but also the way in which "monologic" authority is being actively undermined by Melville. As Spanos explains, "Ishmael's distrust of gravity (high seriousness) or, conversely, his irreverent levity, is not a casual impropriety; it is, rather, a deliberate means of preserving the free play of his mind against the imperial imperatives of logocentric structure."[44] Far from "silencing" other voices or shutting down the conversation, this parodic narrator calls into question the very "monologic authority" with which Moretti is so concerned. Moralizing, arrogant, or smug as the concluding remarks to the cetological chapters may be, they open the discussion to a wide range of possible interlocutors, even if these other personae be masked.

Ishmael as narrator, as character and as symbol, has long enabled an interpretation of *Moby-Dick* that encourages the transformation of Melville's literary narrative into an American national narrative. My reading of "Etymology" and "Extracts," which takes these chapters *as Ansatzpunkt* counterposed to the "Call me Ishmael" point of departure, provides an alternative interpretation that pries the novel loose from a dominant meta-narrative of twentieth-century American Studies, and—with much more to be said—enables a recasting of American literature outside of any particularly nationalist project. As discussed in Chapter 3 above, far from being the representative national text, *Moby-Dick* is a perfect example of a world text, whose "geographical frame of reference is no longer the nation-state, but [. . .] the world-system as a whole."[45] *Moby-Dick* combines a critique of the American national narrative with a projection of a baroque cartography of a world system. In Deleuzian terms, *Moby-Dick* combats the State philosophy of national narrative with nomad thought revealed in a postnational narrative hitherto underestimated or unseen by most literary critics in American Studies. Approaching *Moby-Dick* from an Anti-Ishmael perspective opens up space for a new understanding of the novel and of its effects.

Chapter 5

Marine Nomadology: Melville's Antinomy of Pure Reason

Gilles Deleuze's famous distinction between nomad thought and State philosophy finds an unexpected precursor in an observation by the foremost philosopher in the United States in the mid-nineteenth century. In his essay on "History," Ralph Waldo Emerson describes the same antagonism as the very basis of civilization:

> In the early history of Asia and Africa, Nomadism and Agriculture are the two antagonist facts. The geography of Asia and of Africa necessitated the nomadic life. But the nomads were the terror of all those whom the soil or the advantages of a market had induced to build towns. Agriculture was therefore a religious injunction, because of the perils of the state from nomadism. And in these late and civil countries of England and America those propensities still fight out the old battle, in the nation and in the individual.[1]

This conflict, figured forth in the guise of the nomadic wanderer and the earth-bound, *static* settler, finds a resonant force in Melville's unsettled and unsettling narrative in *Moby-Dick*. Whereas the national narrative of the early and mid-nineteenth century established a sort of State philosophy—and the nationalization of literary narrative in twentieth-century criticism has certainly promulgated a kind of State philosophy in the form of American Studies—Melville's work confounds the process. The conflict between Melville's nomad thought and national narrative's State philosophy into whose service *Moby-Dick* has been put amounts to an antinomic conflict, an antagonistic battle between two opposed sets of laws governing thought and art. Melville comes down strongly on the side of the nomads, and his nomad thought constituted a forceful aspect of his critique of national narrative or State philosophy. The postnational *Moby-Dick* revealed in an anti-Ishmael approach to the novel also explores the antinomies of national narrative in projecting an image of the world system.

The antinomic conflict in *Moby-Dick*, between the statist tendencies toward national narrative on the one hand and those which explode nationalism in favor of a nomadic exploration of a postnational world system on the other, plays itself out on the same intellectual battlefield as that on which Emerson's agricultural and mercantile statists struggle with the perilous nomads. *Moby-Dick,* with its "meanest mariners, and renegades, and castaways"—*Isolatoes,* "not acknowledging the common continent of men, but each *Isolato* living on a separate continent of his own"[2]—provides a fitting cast of nomads for the drama of the "circumnavigating Pequod" on its nomadic trajectory. But given the metaphorical significance of the antinomic "Ship of State," some critics have directly associated Melville's voyage with the United States' own national narrative. For example, Alan Heimert has famously argued that *Moby-Dick* is a political allegory of the impending national crisis in the years leading up to the Civil War, going so far as to assign regional significance to the mates and harpooners (e.g. Flask represents the South, whose fortunes are supported on the backs of African labor, just as Flask climbs atop Daggoo's broad shoulders) and associating Ahab with the secessionist John C. Calhoun.[3] If Heimert's argument is to be believed, Melville's is national narrative, one that is clearly a narrative of national tragedy, not enunciated in the optative mood of the American Renaissance, but in the form of a jeremiad. If the Pequod be a ship of state, Melville certainly does not send it on a voyage of optimistic promise, where it might shine forth as a beacon to the rest of the world. Melville does not produce a city on a hill but a world-circling ship on a voyage of destruction. However, in establishing a narrative of national destruction, Melville is offering a critique of the national narrative itself. Melville's literary cartography contains the national, in more ways than one, even as it insists on presenting a postnational world system. The nomad thought of *Moby-Dick* wrestles with the State philosophy it encounters, highlighting the irreconcilable differences between the two ways of understanding and mapping the world.

Conflicting Forces

Although the terms are most famously employed in his collaborative work with Félix Guattari, Deleuze first makes the distinction between State philosophy and nomad thought in his discussion of the baroque conflict between Descartes and Spinoza. In *Difference and Repetition,* Deleuze associates the "nomadic distribution" of the various components of Being in Spinoza, opposing it to the Cartesian theory of substances that, like the agricultural or statist model, distributes by dividing into fixed categories, demarcating territories and fencing them off. Hence, even before the *Anti-Oedipus* and *A Thousand Plateaus,* Deleuze understood the task of thinking in terms of space, territory, deterritorialization, and reterritorialization. Deleuze notes that the statist or Cartesian distribution of Being is rooted to agricultural need to set proprietary boundaries and fix territories. Alternatively, there is "a completely other distribution, which must be called nomadic, a nomad *nomos*, without property, enclosure or measure" that does not involve "a division of that which is distributed but rather a division among those who distribute themselves in an open space—a space which is unlimited, or at least without precise limits."[4] Famously, Deleuze will go on to associate the thinkers whose philosophies proceed according to this view of Being as his "nomads," themselves distributed throughout the history of philosophy while also standing somewhat outside of it: "I liked writers who seemed to be a part of the history of philosophy, but who escaped from it in one respect, or altogether: Lucretius, Spinoza, Hume, Nietzsche, Bergson."[5] Although Deleuze is discussing the institution of philosophy in particular, one is tempted to add Melville to the list, which Deleuze essentially does through his interpretation of *Moby-Dick* and other works.[6] For Deleuze, these thinkers stand outside of a philosophy which has ever associated itself with the State. "For thought borrows its properly philosophical image from the state as beautiful, substantial or subjective interiority [. . .] Philosophy is shot through with the project of becoming the official language of a Pure State." Deleuze most often cites Descartes and Hegel as State philosophers *par excellence*, but it seems to me that the figure he wrestles with most strenuously is Immanuel Kant, the ultimate philosopher of the Enlightenment whose exploration of the antinomies of pure reason helped to establish, both theoretically and in explicit terms, the antagonism between State philosophy and nomad thought.

In the *Critique of Pure Reason*, a work that exerted immense influence over the Romantics and, by extension and filtered through them, Melville himself,[7] Kant pointedly and with approval identifies metaphysics, "the Queen of all the sciences," with the State. In his preface to the 1781 (first) edition, Kant writes that the threat to this Queen's beneficent government lay in the forces of the skeptics, "a species of nomads, despising all modes of settled life, [who] broke up from time to time all civil society. Happily they were few in number, and were unable to prevent its being established ever anew."[8] Of course, Kant is also critical of the dogmatists, under whose administration the Queen's government was "despotic," which in part led to the "complete anarchy" that allowed such nomads to breach the walls of the kingdom. Kant's metaphor establishes the conflicting philosophical traditions explicitly as statist on the one hand and nomadic (anti-statist) on the other. In the antinomy of pure reason, to which Kant devotes the center and the bulk of the *Critique of Pure Reason*, this intellectual warfare plays itself out. The antinomy reveals the questions that human reason must ask but cannot definitively answer, dialectically opposed but irreconcilable positions, stated in the form of *thesis* and *antithesis*. The battle lines are drawn between parties to the dispute as well: the dogmatic statists who affirm the truth of the theses and the skeptical nomads who affirm the truth of the antitheses. Notwithstanding the irreconcilability of the positions, it becomes clear that Kant comes down in favor of the theses as well, not by virtue of pure reason but with respect to "the interests of humanity."[9] Not surprisingly, then, the argument of the theses is consistent with the aims of State philosophy or national narrative, whereas the antitheses represent convictions of nomad thought.

Kant's analysis of the antinomies of pure reason is lengthy, but he summarizes succinctly the four ("neither more nor fewer") basic cosmological questions raised:

> Whether the world has a beginning [in time] and any limit to its extension in space; whether there is anywhere, and perhaps in my thinking self, an indivisible and indestructible unity, or nothing but what is divisible and transitory; whether I am free in my actions or, like other beings, am led by the hand of nature and of fate; whether finally there is a supreme cause of the world, or whether things of nature and their order must as the ultimate object terminate thought—an object that even in our speculations can never be transcended: these are the questions for the solution of which the mathematician would gladly exchange the whole of his science.[10]

In Kant's argument pure reason is ineluctably led to these unanswerable questions, and neither side of the arguments—for or against the world's finitude, for or against this existence of a soul, for or against the existence of free will, for or against the existence of God—can achieve a definitive, lasting victory. Indeed, in presenting these antinomies, Kant deems himself, and advises the readers to be, an impartial umpire or judge. Nevertheless, although pure reason cannot decide the argument, "for the practical interests of humanity," we do indeed take sides. "[I]f we consulted only our interest and not the logical criterion of truth," we would be able "to comprehend why the participants in this quarrel [. . .] have preferred to fight on one side rather than the other."[11] The "dialectical battlefield" also sets forth the cosmological foundations for the Deleuzian distinction between State philosophy and nomad thought, or—as I argue—the distinction between the statist American national narrative into which *Moby-Dick* has been placed by critical tradition and the nomadic, proleptically postnational narrative Melville produces.

In *Moby-Dick*, Melville comes down firmly on the side of the nomads, favoring the arguments of the antitheses rather than the theses, thus joining the forces of the empiricists or skeptics (as Kant names that side). It is tempting to add, given the initial reception of the novel and Melville's subsequent career, that Melville's position does not have the "advantage of popularity" possessed by the rationalist theses. In the conflict between the theses and the antitheses, between State philosophy and nomad thought, as Kant observed, "the world hails the one with eager approval, and is implacably prejudiced against the other." Kant actually concedes that the empiricist arguments tend to be more sound and more philosophically convincing than those of the dogmatists in support of the theses, but he insists that "*empiricism* is entirely devoid of the popularity of transcendentally idealizing reason" and thus cannot "acquire any considerable influence in the general life or any real favor among the multitude."[12]

Moby-Dick's relative unpopularity is not what makes it an exemplary postnational text, but its tepid sales do suggest that it was somewhat out of touch with the principal concerns of the nation in its time (especially given the tremendous popularity of *Uncle Tom's Cabin* that same year). In our time, after the Melville Revival and hypercanonization of *Moby-Dick,* one might say that the novel has been retroactively renationalized. As noted in the last chapter, the Ishmael-centric interpretations of *Moby-Dick* have shaped the novel in such as way as to fit neatly into the image repertoire of American national narrative, consisting of an American Adam, a Virgin Land, and an errand into the wilderness. Donald Pease's definition of the term explicitly connects national narrative to State philosophy and to Kant's own project. Referring to the conflict between a universal, integrated national identity and the various forces within societies that resist incorporation into the identity, Pease describes the Enlightenment origins of national narrative:

> This contestation between social demands developed by political movements and a nation-state's power to misrecognize them [. . .] originated from within the discourse of the Enlightenment. The term *national narrative* itself refers to the process whereby the discourse of Enlightenment produced particulars—nation-states—out of universal norms: Reason, Equality, Social Justice, Liberty. Acting as agents of the state, these national narratives constructed imaginary relations to the actual sociopolitical conditions to effect imagined communities called national peoples.[13]

In other words, national narrative acts as a State philosophy to establish and to bolster the idea of the nation itself. Even in the seemingly apolitical and ahistorical antinomy of pure reason, the State philosophy or nomad thought competing on the "dialectical battlefield" functions, respectively, to establish or thwart a national narrative. In the specifically *American* national narrative, the image repertoire assumes a victory for the dogmatic theses, and the dogmatic, Ishmael-centric reading of *Moby-Dick* champions the cause. With a discrete, Adamic Ishmael functioning as an individual subject imbued with free will, operating in a spatiotemporally finite world on an errand into the wilderness in search of a Promised Land established by Divine Providence, *Moby-Dick* can be manipulated into a national narrative conforming to the State philosophy of Kant's dogmatic theses. However, in my view, this involves misreading *Moby-Dick* and Melville's postnational project.

Moby-Dick addresses the antinomy of pure reason, recast in terms of an American national narrative and the resistance to it, throughout the body of the novel. For Melville, the State philosophy propounded in the theses is not merely that of the theo-logocentric tradition of the Puritans and the American

Jeremiad (as William Spanos would have it), but also that of the anthropo-logocentrism of mid-nineteenth-century, Emersonian humanism. "Manifest Destiny," in whose overall ideological project Melville has been implicated by a number of critics,[14] labels the shift in American cultural history from the theo-logocentric to the anthropo-logocentric, insofar as manifest destiny maintains the divinely ordained "errand into the wilderness" while supplementing the religious injunction with a humanism that demands individual agency in the movement through, and conquest of, geographical spaces. That is, while the force of the Puritan *logos* retains no small authority, particularly in the national imagination that can reiterate and ritualize the Puritan errand in its political discourse, it is really a *residual* form, no longer the dominant force. The Puritan *logos* still exists alongside of, and lends religious sanction to, the *dominant* anthropo-logocentric social formation in which territorial expansion and politico-economic centralization (the centripetal and centrifugal forces of the nation-state) are combined to constitute the American national project. *Moby-Dick,* in mapping the *emergent* postnational world system, envisions a space in which this national formation is itself shown to be but a transitory moment, rather than a destination, goal, or norm. The nomad thought of *Moby-Dick* thus challenges the State philosophy of American national narrative.

First Antinomy: "Saturn's Grey Chaos"

In Melville's treatment of the first antinomy, the battle over whether the world is finite or infinite in time and space, the conflict can be dramatized by the contesting images of man and whale, and more generally of land and sea. The whale itself, that inscrutable and *inhuman* force, is set up against the human, all-too-human, which will be found wanting. And *the* whale, Moby Dick, becomes the apotheosis of this force, the very emblem of infinitude, for infinite extension over time and space. Melville uses the figures of the sea, the whale, and *the* whale Moby Dick as avatars of the first antithesis in the antinomy, opposing them to the finitude of land and man in an effort to undermine a reigning State philosophy of the American Renaissance. In American national narrative, the inevitability of human advancement folds into the national errand into the wilderness, spoken in an optative mood and undertaken with the certainty of the faithful; an ineluctable geographical and historical advance thus figures in a national image of permanence and futurity. This involves a finitude inasmuch as the spatiotemporal world of national narrative must be achievable, with a visible end. *Moby-Dick* mocks this narrative, posing the whale's utterly destructive force as an insurmountable opponent, who taunts "baby man"[15] and his infantile sense of self-importance. The whale's infinitude is set against man's finitude, and the baroque complexity of the nomad thought undermines the rigid, administrative regularity of State philosophy.

Throughout *Moby-Dick*, the dogmatic thesis of national narrative falls under almost constant attack. In "The Fossil Whale," not only is the whale immortal, existing across and through time, but the whale exists *before* what man calls time. The "pre-adamite" whale is older than the dinosaurs, "ere time itself can be said to have begun; for time began with man." The whale is beyond human time, which must have a discrete beginning and a foreseeable end, a Genesis and an Armageddon. But the "antechronical Leviathans [. . .] having been before all time, must needs exist after all humane ages are over." The narrator of "The Fossil Whale" is overcome by the sheer weightiness of contemplating such temporal infinitude. "Here Saturn's grey chaos rolls over me [. . .] I am horror-struck at this antemosaic, unsourced existence of the unspeakable terrors of the whale."[16] The whale too is a figure of grey chaos, *grey* in color, but also in that mysterious greyness that makes things obscure, ambiguous, or—in Ahab's apt term—"inscrutable." The black-and-white-at-the-same-time greyness does away with black and with white, making what had seemed to be legible signs into scarcely intelligible figures, rendering order into chaos, and overwhelming a national narrative of fixed meanings and representative subjects. As noted in the last chapter, *overwhelm* literally means "rolls over," which is the etymological origin of the *whale* itself. The whale is associated with destruction and chaos throughout *Moby-Dick*, but "Etymology" had already told us that the very word indicates its *overwhelming* nature, as "the whale" is associated with "roundness and rolling." In both its individuality and its species-being, the "pre-adamite" whale *rolls over* the American Adam.[17]

Saturn, or Chronos, lord of the titans and father of the gods, is an apt representative of the whale's infinitude. Not only does the titan stand for time itself, but for the tremendous and overwhelming force that Melville associates with the whale. Antedating not only men but also the gods themselves, Saturn extends infinitely back and infinitely forward; such a figure is beyond any human destiny, manifest or otherwise, and stands in stark contrast to the hopeful, self-confident, and largely triumphalist American Adam of national narrative. In *Moby-Dick*, the human destiny is not made manifest in the progressive march of time except that it marches toward death and destruction, for the human to be overwhelmed by the inhuman and infinite force of the whale. Whereas man is finite, the whale began before time itself and will continue infinitely, in Melville's view. Melville, in naming the great monuments to supposedly "lasting" nation-states, specifically derides national self-pride and hubris in his discussion of the whale's immortality and, by extension, of mankind's relatively short-lived mortality. Although some "philosophers of the forecastle," asserting their own false claims to overwhelm the whale, might argue otherwise, Melville insists that "we account the whale immortal [. . .] He swam the seas before the continents broke water; he once swam over the sites of the Tuileries, and Windsor Castle, and the Kremlin. In Noah's flood he despised Noah's Ark; and if the world is to be again flooded, like the Netherlands, to kill off its rats, then the eternal whale will still survive, and rearing upon the topmost crest of the equatorial flood, spout his frothed defiance to the skies."[18] The infinitude of the whale, the avatar of the first antithesis in Melville's own nomad thought, likewise defies the aspirations of national narrative.

Moby Dick himself is a figure of infinitude. In chapter after chapter we learn of the legendary extent of the whale. The legends merely enlarge upon the subject of legend, just as "The Whiteness of the Whale" posits the infinite meanings of that whale in the semantic universe. In the eponymous chapter, "Moby Dick," we are told that this particular whale is "ubiquitous; that he had actually been encountered in opposite latitudes at one and the same instant of time." A few lines later, we learn that "Moby Dick [is] not only ubiquitous, but immortal (for immortality is but ubiquity in time)."[19] Moby Dick's infinitude, his limitlessness in time and space, combined with his well known though indeterminate bodily magnitude and force, contrast to the relatively finite and impotent human condition. And, in the final conflict between Moby Dick and the *Pequod*, we are not surprised to see the finite man (perhaps even the ship of state, as some critics argue) overwhelmed by the infinite white whale.

The point is underscored in Melville's ironic but apt contrast between land and sea, each with its attendant imagery and effects. As Melville is well aware, American national narrative is very much "of the land," not merely in its representative ideological locale of the Virgin Land, but in its overall tone and attitude. Notwithstanding its maritime beginnings in the transatlantic voyages of Columbus, Walter Raleigh, the Jamestown settlers, or the Mayflower Pilgrims, national narrative partakes of the soil, charting territories conquered and settled, celebrating pioneering woodsmen like Natty Bumppo and Davy Crockett who clear the paths and protect the hearths of new homesteaders, establishing the images of the yeoman farmer and the town meeting as the representative individual and political models.[20] Implacable nature is something to be tamed, to be overcome, and the triumph over such nature gradually moves from fable and folklore to deeply felt history, becoming the story of a person, a people, a nation. These audible themes of national narrative intertwine into a landlubber's symphony in which one really might actually view one's destiny as self-directed, as lying entirely within one's own hands. In land-based narrative, attitudes develop: if something gets in your way, knock it down, dig it up, shoot it. The landlubber is, or quickly becomes, arrogant. On this land, the representative American national narrative—the story of an individual American Adam making his way in a Virgin Land on a divinely ordained errand into the wilderness, destined to be triumphantly successful and a model for the entire world—practically writes itself.

The sea provides the foundations for different stories, stories much more difficult to assimilate into national narrative. At sea, nature cannot be conquered or tamed, but must be treated with the utmost respect if man is to survive the encounter. Wind and waves can be negotiated or navigated (literally), but they cannot be conquered or mastered. Different attitudes develop: if something gets in your way, change direction, trim sail, emend plans, alter expectations. The seaman requires the very flexibility that the landlubber proudly abandons. Whereas the landsman might swaggeringly describe obstacles conquered, the sailor humbly recounts ordeals survived. If, according to a certain reading of *Moby-Dick*, Ahab's arrogance with respect to Nature leads to his downfall, it might be viewed as a case of Ahab as the American Adam attempting to impose his Virgin Land mission on an unwilling and unforgiving ocean. In any event, the mastery of the sea is a far cry from the domination of a territory, and even Ahab's prodigious whaling skills are no match for the antechronical Leviathans.

In Melville's nomad thought, which despises "all modes of settled thinking," the sea tauntingly undermines human and American hubris. The examples are too numerous to recount, but a passage from the chapter "Brit" is representative of Melville's argument.

> [A] moment's consideration will teach, that however baby man may brag of his science and skill, and however much, in a flattering future, that science and skill may augment; yet for ever and for ever, to the crack of doom, the sea will insult and murder him, and pulverize the stateliest, stiffest frigate he can make [. . .]
>
> The first boat we read of, floated on an ocean, that with Portuguese vengeance had whelmed a whole world without leaving so much as a widow. The same ocean rolls now; that same ocean destroyed the wrecked ships of last year. Yea, foolish mortals, Noah's flood is not yet subsided; two thirds of this fair world it yet covers. [. . .]
>
> No mercy, no power but its own controls it. Panting and snorting like a mad battle steed that has lost its rider, the masterless ocean overruns the globe.

Howsoever man may take pride in terrestrial conquest, the sea's power—like the whale's—always overwhelms him. In the same section Melville directly implicates American national narrative in this critique by including a swipe at one of its seminal figures, Christopher Columbus: "we know the sea to be an everlasting terra incognita, so that Columbus sailed over numberless unknown worlds to discover his one superficial western one."[21] Thus, Melville takes the Virgin Land and the Providential American *topos* to be a mere "superficial western world," insignificant in comparison to numberless worlds still unknown. Just as "the sea rolled on as it rolled five thousand years ago" in *Moby-Dick*'s closing lines,[22] the human and finite world is overwhelmed by the oceanic, infinite world. In its map of this world system, *Moby-Dick* lays bare the pretensions of national narrative.

The reconfiguration of the first antinomy of pure reason in terms of man and whale, land and sea, picks apart the hopeful and celebratory national narrative by emphasizing its limits (and limitedness) and undermine its own arguments. The "green, gentle, and most docile" Virgin Land is perpetually surrounded by the "appalling ocean,"[23] whose force cannot but overwhelm, roll over, its settled and settling assumptions. The infinitude of the oceanic whale is directly contrasted to the finite, terrestrial man, and this antagonism underscores the faulty logic of the American national narrative's State philosophy. Melville's ironic and pointed examples thus reveal the discourse of "manifest destiny" to be hollow.

Second and Third Antinomy: "Is Ahab, Ahab?"

The second and third antinomies, dealing with the question of whether there is a simple substance or soul and whether such a soul has free will, may be dealt with jointly, since the latter enfolds the former in establishing a subject with agency, or a free-willing subject. *Moby-Dick*'s underlying critique of national narrative involves a critique of the foundational individualism inherent in the American Adam and, as Alexis de Tocqueville noted, in American culture at large. Individualism, in theory, combines the theses of the second and third antinomies by insisting that the individual subject is a pure and simple substance, an identifiable unity in itself, with the innate ability to act. Tocqueville, who coined the term *individualism* in *Democracy in America*, was amazed by the force of the conviction in the minds of the Americans he observed. Tocqueville's key conceptual category for the work was not individualism, but rather *equality*, but he shows how individualism is the direct result of the latter.

> As social conditions become more equal, the number of persons increases who, although they are neither rich nor powerful enough to exercise any great influence over their fellows, have nevertheless acquired or retained sufficient education or fortune to satisfy their own wants. They owe nothing to any man, they expect nothing from any man; they acquire the habit of always considering themselves as standing alone, and they are apt to imagine that their whole destiny lies in their own hands.[24]

Tocqueville adds that, contrary to its apparent ideals or to one's expectations, "Democracy leads men not to draw near their fellow creatures; but democratic revolutions lead them to shun each other and perpetuate in a state of equality the animosities that the state of inequality created."[25] Pure individualism, like Hobbes's state of nature, results in a *bellum omnium contra omnes*; since all are equal and each feels self-sufficient, a war of all against all is inevitable. Community, including the imaginary community, would seem to suffer in light of this ideology. Hence the great irony and paradox of America: individualism, a sentiment that expressly disavows collective or communitarian ideals in favor of assiduously asserted individual will, becomes the ideological cornerstone of the distinctively American identity, the identifiable feature of the collective national community.

Melville's nomad thought assails this key element of American State philosophy by calling into question both aspects of the individual agent—that is, both the integrity of individuality and the existence of free will or agency. Obviously, individualism requires an individual, a soul or ego unique to each person, who can and does *act*. Agency is thus inseparable from the agent. Individualism is essential to the State philosophy of American national narrative; however much the national mission (errand into the wilderness) may be divinely ordained, destiny made manifest, the Adamic American individual must be the agent of that destiny. Like those persons Tocqueville encountered in his travels, Americans "imagine their whole destiny to be in their own hands." Melville calls such agency into question again and again in *Moby-Dick*, yet most readers hew to an interpretation of the novel that has become dogmatic. This interpretation ignores Melville's thoroughgoing critique of the concept of agency, and recasts the plot of the novel as a battle of wills—Ahab's versus Ishmael's, or Ahab's versus Starbuck's, and so on. Such a reading of *Moby-Dick*, I have argued, follows from taking Ishmael to be the central figure of the novel, and necessarily ignores one of the most powerful lines of thought, of nomad thought, cutting through the entire novel: the polemic against a free-willing subject.

Famously, in "Loomings," the narrator refers to "the invisible police officer of the Fates, who has constant surveillance of me, and secretly dogs me, and influences me in some unaccountable way," who somehow led him to a whaling voyage rather than a merchant ship. The "Ishmael" of this chapter concedes that even his belief in his own free will was a delusion, supplied by "those stage managers, the Fates [. . .] cajoling me into the delusion of that it was a choice resulting from my own unbiased freewill and discriminating judgment."[26] The idea of police surveillance is intriguing and paradoxical, of course, inasmuch as it suggests the illogical proposition that fate may not be ineluctable. Surely the Fates do not need police officers to enforce the laws of fate, and one fated to do something cannot *not* do it; Oedipus and the entire tradition of Ancient Greece testify to the inescapability of fate. But Melville here does not describe the police officers of the Fates as enforcing the law, catching and punishing those who break it, but rather he notes the way in which the officers' "constant surveillance" affects how the individual acts in advance of breaking any laws. Like Michel Foucault's analysis of panopticism in *Discipline and Punish*, Melville is suggesting the ways in which behavior—thought to be a product of one's own free will, but discovered to be otherwise—is conditioned, influenced "in some unaccountable way," by surveillance. By calling into question the narrator's agency, and by noting that the narrator-observer who "sees" what happens is himself observed, Melville deliberately muddies the waters of the narration of

Moby-Dick. Ishmael, the very model of the American Adam in a dogmatic reading of the novel, himself lacks the individual agency of the representative subject of national narrative. As a character, a participant-observer who narrates the story, this Ishmael-figure cannot be the American Adam whose destiny is manifestly in his own hands; his free will is a delusion, and his actions (including, presumably, his own narration) is conditioned by the Fates' policemen.[27]

Later, when "choosing" which ship to join, one such police officer of the Fates is named. Queequeg's "little black god," Yojo, "earnestly enjoined that the choice of the selection of the ship should rest wholly with me [. . .] [Yojo] had already pitched upon a vessel, which, if left to myself, I, Ishmael, should infallibly light upon, for all the world as though it had turned out by chance." Here again, Ishmael indicates that his free will is illusory, inasmuch as Yojo has previously determined his selections. But Ishmael will equivocate, suggesting a few lines further down that Yojo "meant well," but "in all cases did not succeed in his benevolent designs."[28] Thus, here, Ishmael directly questions the power of a sort of fate over the free-willing subject, whereas in "Loomings" he had only implied that fate or determination might hold sway over him (i.e. by requiring the Fates to have a policeman who could keep the free-willing subject "in line"). Notwithstanding the frequent allusions to the rule of fate, necessity, or chance over an illusory free will, Ishmael continues to assert his freedom, only to find himself again and again subject to powers well beyond his own will.

A most striking example of this can be found in "The Mat-Maker." This chapter might be noteworthy for its attempted resolution of the antinomic conflict, insofar as Ishmael (the first-person narration, with its "I," returns here following a chapter ["surmises"] in which the "I" was attributed to Ahab) claims that "chance, free will, and necessity" are "no wise incompatible—all interweavingly working together." Ishmael also asserts his personal sovereignty over the countervailing power of fate or necessity. In language unmistakably resonant with Tocqueville's characterization of individualism, Ishmael thinks to himself, "with my own hand I ply my own shuttle and weave my own destiny into these unalterable threads." But Ishmael also takes the figure of the "straight warp of necessity" to show the limits of his will, and concedes that chance, here figure in Queequeg's "impulsive, indifferent sword," rules either of them at turns, "and has the last featuring blow at events."[29] Whereas an American individualist appropriates for himself not only a freedom to resist fate, necessity or history (or "the Loom of Time") but also a freedom to determine fate itself—in other words, manifest destiny—Ishmael, in claiming that his free will is "prescribed" by the dual forces of necessity and chance, dials down the hubris of individualism. Elsewhere, in allowing that his free will is illusory, Ishmael, like his successor conceptual personae, Pierre and Bartleby, becomes a figure of that ambiguity which pervades Melville's corpus. The uncertainty of one's own unitary self and its ability to act is a pervasive theme in *Moby-Dick*.

As noted in the last chapter, Ishmael himself is not really a unitary figure, as he is (or he adopts) a conceptual persona as befits the narration at different times. The problem of narrative voice in *Moby-Dick* has led to the establishment, in Melville studies, of a dual subject, "Melville/Ishmael," an identification which, I believe, is wrongheaded. For instance, Howard Vincent speaks of Melville "becoming" Ishmael in the opening pages of *The Trying-out of Moby-Dick*.[30] Ishmael-centric readings of the novel mostly take for granted that the author, Melville, speaks in the voice of Ishmael, which aids in rendering the novel a sort of personal and then national narrative. This identification of Melville and Ishmael relates to the problem of individuality and free will (i.e. to the second and third antinomies of pure reason), and leads to further confusions.

Even when a critic does not explicitly identify Melville with Ishmael, the Ishmael-centric reading often persists. For example, in Dimock's influential study of Melville's "poetics of individualism," Ishmael is hardly mentioned by name at all. Rather than addressing the degree to which Ishmael might be a figure for Melville's critique of American national ideology, Dimock makes the case for Melville himself as the embodiment and proponent of such an ideology. Dimock's argument is rooted in her sense of Melville's self-aggrandizing authority as author. Nevertheless, little would change substantially in her critique if the Ishmael's name were substituted for Melville's, if Ishmael were the "whale author" of the "imperial folio."[31] In Dimock's argument, the author (Melville) assigns his characters, especially Ahab, a freedom precisely in order to doom them:

> . . . freedom is primarily a disciplinary postulate, something that the author imputes to his characters and that enables him to judge them, as free agents, responsible and punishable for their deeds. Freedom of this sort points to a negative individualism: one that produces individuals as "subjects," figures whose very freedom constitutes the ground for discipline.[32]

Thus, for Dimock, Melville is to blame for "blaming the victim," since Ahab is a victim of Melville's own authorial discipline. Negative individualism posits a free-willing individual only in order to ascribe guilt to that person.

Without naming him directly, Dimock invokes Nietzsche and his polemic against "slave morality" (which, after Deleuze, may be just another code word for State philosophy). In *The Twilight of the Idols*, Nietzsche states his position succinctly: "the doctrine of the will has been invented essentially for the purpose of punishment, that is, because one wanted to impute guilt."[33] This insight had been developed at length in *On the Genealogy of Morals* (especially in the second essay), in which Nietzsche discusses the emergence of the "sovereign individual," endowed with free will and hence "responsible" for his actions.[34] Such also is the theoretical and political thrust of Foucault's *Discipline and Punish*, in which the modern system of power (and particularly the power relations involved in surveillance and punishment) creates the individual. The individual is, in effect, born through punishment.[35]

Dimock's argument takes Ahab to be such an individual, a sovereign subject, "by definition a free agent," and thus implicates Melville in a particularly malignant form of national narrative (or State philosophy) that not only serves to establish, bolster, and celebrate the nation-state, but actively seeks to eliminate, rather than incorporate, the forces that do not fit within its integrated whole. Individualism, the embodiment of the second and third theses in the antinomy of pure reason combined (i.e. of a discrete ego that has free will) as well as the fundamental ideological cornerstone of American national narrative, becomes the tool to accomplish this project.

> To dispatch Ahab, to disarm him, to make him die not only inevitably but also deservedly, Melville needs an executory instrument, a logic that explains and justifies the fate of this character. That logic is all too easy to come by, as it is already a provision in individualism.[36]

Ahab must be both doomed (i.e. fated to die) and responsible (i.e. free to choose a course of action that will require his own death penalty). Dimock notes how similar certain characterizations of Ahab are to those of contemporary discussions of Indians, who are also considered simultaneously doomed and free, and she suggests that both Ahab and Indians "are generated by a common representational form of the self, one that produces not only literary meanings but also social meanings, and not only fictive casualties but also casualties of flesh and blood."[37] Thus, for Dimock, through the authorial, disciplinary form of free will, Melville establishes a national narrative that directly corresponds to the most vicious and violent aspects of manifest destiny as imagined and as practiced in the mid-nineteenth-century United States.

However, the basis for Dimock's argument, its *sine qua non* in fact, is far from certain. Whereas Dimock takes for granted that Ahab is "by definition" a free agent, the text of *Moby-Dick* frequently calls such a notion into question. Ahab may not be a "free agent" at all. In "'A Romantic Book': *Moby-Dick* and Novel Agency," Jonathan Arac provides a careful reading of "The Chart" in order to demonstrate the degree to which Ahab's freedom is formulated as problematic, and certainly not established as a unquestionable fact, in *Moby-Dick*. Arac's reading reveals a troubled and troubling sense of agency, where Ahab may not have the iron will that many readers assume he has or that Dimock imputes is granted to him (in order then to punish him) by Melville. In the penultimate paragraph of "The Chart," one can see Ahab's agency and subjectivity flickering in and out of existence, right at the sentence level, and the effect is to render dubious the very will and ego usually attributed to him.

> But granting all this; yet, regarded discreetly and coolly, seems it not but a mad idea, this; that in the broad boundless ocean, one solitary whale, even if encountered, should be thought capable of individual recognition from his hunter, even as a white-bearded Mufti in the thronged thoroughfares of Constantinople? Yes. For the peculiar snow-white brow of Moby Dick, and his snow-white hump, could not but be unmistakable. And have I not tallied the whale, Ahab would mutter to himself, as after poring over his charts till long after midnight he would throw himself back in reveries—tallied him, and shall he escape? His broad fins are bored, and scalloped out like a lost sheep's ear! And here, his mad mind would run on in a breathless race; till a weariness and faintness of pondering came over him; and in the open air of the deck he would seek to recover his strength. Ah, God! what trances of torments does that man endure who is consumed with one unachieved revengeful desire. He sleeps with clenched hands; and wakes with his own bloody nails in his palms.[38]

At the beginning of the paragraph, the narrative voice is the familiar third-person, although perhaps with an Ishmael-like figure doing the discoursing,[39] but what Arac calls a "trans-subjective agency" is implied in the sentence that starts "And have I not tallied the whale." The personal pronoun, "I," here refers to Ahab, to Ahab's own thoughts or narrative, yet without any quotation marks to distinguish his discourse from the narrator's "Ahab would mutter to *him*self" (where the third-person narration reappears); then the sentence returns to Ahab's point of view after the dash, such that the "him" now refers to Moby Dick, not to Ahab. The subject literally slips, and the individual ego is not so easy to pin down. This trans-subjective agency, at the level of textual practice, "erodes the firm contours of individuality so important to Dimock's argument."[40] Dimock insists that Ahab's fixed identity must be assured, that fixing Ahab's identity is Melville's strategy of containment. She maintains, for instance, that the line, "Ahab is for ever Ahab," "must stand as the very epigraph of negative individualism, of the punitive logic centered on the autonomous self."[41] But that line is itself evidence of Ahab's lack of will, a sign that he is "the Fates' lieutenant; I act under orders." Immediately after stating that "Ahab is for ever Ahab," the captain adds, "This whole act's immutably decreed." Furthermore, Dimock entirely overlooks another, earlier conversation between Ahab and Starbuck in which Ahab asks, "Is Ahab Ahab? Is it I, God, or who, that lifts this arm?"[42] This cosmic question undermines Dimock's assertion that Ahab possesses sovereign, though negative, individuality. Here, as elsewhere, Ahab himself questions his own identity and his own agency. The power of necessity, a physical force such as that which makes stars revolve (as Ahab points out in the next line), is at odds with the notion of a clear agency embodied in the sovereign individual.

More importantly perhaps, Ahab's free will is directly challenged in the final paragraph of "The Chart," where we see Ahab plagued by dreams that drive him from his state room, "as though escaping from a bed that was on fire." The subsequent passage that concludes the chapter is worth quoting at length:

> For, at such times, crazy Ahab, the scheming, unappeasedly steadfast hunter of the white whale; this Ahab that had gone to his hammock, *was not the agent* that so caused him to burst from it in horror again. The latter was the eternal, living principle or soul in him; and in sleep, being for the time dissociated from the characterizing mind, which at other times employed it for its outward vehicle or agent, it spontaneously sought escape from the scorching contiguity of the frantic thing, of which, for the time, it was no longer an integral. But as the mind does not exist unless leagued with the soul, therefore it must have been that, in Ahab's case, yielding up all his thoughts and fancies to his one supreme purpose; that purpose, by its own sheer inveteracy of will, forced itself against gods and devils into a kind of self-assumed, independent being of its own. Nay, could grimly live and burn, while the common vitality to which it was conjoined, fled horror-stricken from the unbidden and unfathered birth. Therefore, the tormented spirit that glared out of bodily eyes, when *what seemed to be Ahab* rushed from his room, was for the time but a vacated thing, a formless somnambulistic being, a ray of living light, to be sure, but without an object to color, and therefore a blankness it itself. God help thee, old man, thy thoughts have created a creature in thee; and he whose intense thinking thus makes him a Prometheus; a vulture feeds upon that heart for ever; that vulture the very creature he creates.[43]

As Arac notes, this passage clearly divides Ahab, grammatically and otherwise, into a subject and an object, "an active 'agent' and an acted-upon 'him.'" Far from establishing Ahab as a sovereign subject and free agent, the novel actually denies the certainty of unified subjectivity or freedom, a certainty which is essential for the "poetics of individualism" at the crux of Dimock's project, a poetics undergirding American national narrative. Against Dimock, Arac asserts that, "in *Moby-Dick*, individuality is neither a goal nor a premise, but a puzzling possibility."[44]

I have italicized certain phrases in the passage both to highlight elements mobilized in Arac's argument and to introduce a problem rarely discussed directly. Stated briefly, Ahab's quest is not one of vengeance. The final paragraph of "The Chart" suggests this powerfully, as if to undermine the gospel truth that Ahab's "sin," as Starbuck states (calling it "blasphemy") in "The Quarter-Deck," is his "Vengeance on a dumb brute [. . .] that simply smote thee

[Ahab] from blindest instinct."[45] Dimock takes Ahab's vengefulness as another element in the logic of negative individualism ascribed to him by Melville in order to "dispatch him." For, Dimock argues, Ahab is figuratively identified with the American Indians inasmuch as he is both doomed and responsible for his own doom (i.e. a free agent who, paradoxically, chooses his own fate and may therefore be blamed for his choices and actions); Ahab, like the "bad" Indians as characterized by James Fenimore Cooper and Francis Parkman, is also consumed by vengeance. "Vengeance is wrong because it is unprofitable," Starbuck reasons, and it must be repressed since it disrupts the "signifying economy" required for economic and political expansion, figured here as the "Nantucket market."[46] But "The Chart" not only calls into question Ahab's agency and individual subjectivity but his motive of vengeance as well. The actual agent, the supposedly vengeful thing, is not Ahab or even the "living principle or soul" within him; it is rather that purpose, "by its sheer inveteracy of will," which is its own "independent being," the force that moves Ahab. And this purpose, understood heretofore as revenge, cannot in fact be revenge, for its origins lie in an "unbidden and unfathered birth." Contrary to the arguments that assume for Ahab a pure or unregulated individualism, this phrase indicates that neither Ahab, nor the white whale, nor anyone else for that matter has called this "independent being" into being; it is unbidden and unfathered, *causa sui.* The final line of the paragraph reiterates Ahab's duality, as subject and object, in seeking an origin for this purpose, which becomes both something created "in him" and something "he creates." But this appears to be another ruse. The unconditioned (to use Kant's term) existence of this "independent being" cannot be caused by Ahab or by Moby Dick. The source of this purpose cannot therefore be located in some moment at which the white whale dismembers Ahab. If Ahab be vengeful, at least that does not mean that his "fixed purpose" is his vengeance.

Hence, Starbuck's objection to Ahab turns out to be based on a false assumption, as the pious first mate has confused Ahab's motives and personality. Dimock, who had up to that point associated Ishmael's voice with Melville's own, here makes Starbuck the voice of Melville (in her own version of trans-subjective agency, perhaps), identifying Starbuck's objections with Melville's. Famously, in the quarterdeck scene, Starbuck objects that the hunt for Moby Dick is wrong, first because it is bad for business ("How many barrels will thy vengeance yield thee") and second because it is crazy and sinful ("Madness! To be enraged at a dumb thing, Captain Ahab, seems blasphemous")—in that order.[47] Dimock argues that Starbuck, filling in for Melville, longs for the economic order in which things and beliefs circulate within a smooth and harmonious system of exchange. Starbuck stands for stable meanings, for fungible and recognizable units, where the value of a thing may be easily ascertainable, on the Nantucket market *and* in the human heart. Ahab represents a thoroughly disruptive force, not only disrupting the ordinary course of business, but the normal chain of meanings. Ahab is a destructive force like the nomad hoard, transforming the agricultural and grid-like space in which Starbuck and the denizens of the national live, a space of predictable destiny and recognizable regulation. Thus, I believe Dimock is mistaken when she argues that Starbuck and Melville are "logical allies," as Melville clearly presents Starbuck as an ineffectual bureaucrat in the face of a revolutionary and destructive force from the outside. To be sure, Starbuck is also a sympathetic figure (in, for instance, "The Symphony"), but he is in no way the hero of the novel. Notwithstanding the view of Ahab as the resentful man of vengeance, it is Starbuck, with his puritanical Christian ethics and his bourgeois reduction of values to exchange-value, who seems to fit the bill for Nietzsche's man of *ressentiment*, which is to say the soul of vengeance.

Thus the view of Ahab as the man of *ressentiment* is also wrong. Vengeance, as Nietzsche has Zarathustra say, is at the heart of "justice," itself rooted in a discourse of equality.[48] This is but another version of that will to punish, which Nietzsche identifies as the reason that the doctrine of free will was developed in the first place. Starbuck, not Melville (or anyone else on the *Pequod*), blames Ahab. Yet even Starbuck acquiesces—perhaps out of a sense of duty, perhaps out of his own weakness, but more likely because "the ineffable thing has tied me to him"[49]—to the "purpose," which is, after "The Quarter-Deck," no longer Ahab's purpose, but the whole crew's. An "ungodly, god-like man," Ahab himself is beyond good and evil.

Melville's critique of individualism, of both the individual subject and the agency associated with it, forms the basis for his critique of American national narrative as well. In setting himself against the clearly defined, free-willing individual subject, Melville embraces the nomadic energies of another kind of narrative, one that expressly opposes the aims and effects of the national narrative. In the antinomic conflict between State philosophy and nomad thought, Melville helps establish the postnational world he is attempting to map. In establishing the antitheses of the second and third antinomies, Melville strikes at the heart of American State philosophy.

Fourth Antinomy: "Who's to Doom?"

The fourth antinomy, as it is dealt with in *Moby-Dick*, follows directly from these others and may be treated briefly. The matter of God's existence is not the fundamental question; rather how God is imagined defines the problem. Melville's "quarrel with God," and his tortured metaphysics have been the subject of a library's-worth of material,[50] but the key issue in the antinomy of national narrative is how such a godhead is figured. Starbuck's God is the principle of punishment, the disciplinary figure that Dimock conflates with the Author. This God is figuratively allied with the State insofar as one must enter into a (tacit) contractual relationship with Him, as with the social contract, and if one demurs from the terms of that agreement, he will be punished accordingly. Ahab's "blasphemy," as Starbuck understands it, will lead to his damnation, both in the present world and in the hereafter. Ahab's blasphemy is an affront to both the heavens and the social order, since it violates the tacit agreement with God the Father and the signed contract with the owners of the ship, "the business we follow," as Starbuck puts it. A direct connection is established by Starbuck, with his strong Protestant ethic, between religious sin and the disruption of the political economy.

The "voice" that announces Starbuck's moral position, prior even to Starbuck's appearance in the novel, may be found in Father Mapple's sermon.[51] Jonah, in attempting to flee to the ends of the earth from God's "hard command," brings upon himself the punishment of the infernal belly of the "great fish." For Mapple, God must be understood as a punitive force ("O Father—chiefly known to me by Thy rod"), a cosmic policeman who must regulate the mortals under his jurisdiction. Ahab, of course, does not flee from God; more blasphemous, he vies with him. "Talk not to me of blasphemy, man; I'd strike the sun if it insulted me. [. . .] Who's o'er me?" Significantly, Mapple ends his retelling of the Biblical Jonah narrative before Jonah's own quarrel with God (see Jonah iv: 1–11), in which he questions God's decision to punish or refrain from punishing. Ahab is more like the affronted Jonah than the Jonah who Mapple declares is "grateful for punishment."[52]

Ahab's "Who's over me?" challenge appears to be an irreligious and even anti-democratic, because autocratic, dictum. But this is not exactly the case. If anything, Ahab's stance is hyper-democratic. In this "Quarter-Deck" speech, Ahab places himself on equal footing with the sun itself, as he will do later in another conversation with Starbuck (in "The Symphony") in which he speaks of both his and the sun's lack of agency. The equality of Ahab and the sun, as with the multiple connections between Ahab and fire throughout the narrative, points to that "pantheism"—the same pantheism Ishmael on the masthead had warned against, fearing "Descartian vortices"—which animates much of *Moby-Dick*. This pantheism presents a naturalistic God (Spinoza's *Deus sive Natura*), a creator inseparable from creation, and therefore unfit or unable to judge and condemn. As Ahab puts it, "Who's to doom, when the judge himself is dragged to the bar?" This image of pure immanence constitutes the *fröhliche Wissenschaft* of Nietzsche and Spinoza, directly opposed to Starbuck's and Mapple's "metaphysics of the hangman."[53]

Reaction to Spinoza's philosophy indicates clearly that this kind of pantheism may also be understood as atheism, however paradoxical it may seem. But for the theo-logocentrism of the American jeremiad, a God who does not (and cannot) judge and condemn is not really God at all. Ahab's "Who's to doom?" is perhaps the most blasphemous sentiment in the novel under the circumstances. *Moby-Dick* thus poses Ahab's, and frequently the narrator's, pantheism as the antithesis of the dominant Puritan Christianity that ideologically underwrites the American jeremiad and American national narrative. In the somewhat more secular American Renaissance version, the anthropo-logocentric view (as Spanos would have it), the figure of God is replaced by a state or by a self-regulating individualism; either way, the bad conscience of a juridical individualism upholds the ethos of a national narrative than can contain a myth like manifest destiny. Removing the Puritans' God from the secularized American narrative did not remove His function; the State philosophy persists. The pantheism or naturalism, which is also an atheism, of *Moby-Dick* as a whole suggests that the novel takes the side of the antithesis, even if that means the end of America itself, as the sea and its leviathans will roll on as they did ere "baby man" first appeared. Like Nietzsche's immoralist, *Moby-Dick*—voiced by multiple characters—insists that "there is no end. One is necessary, one is a piece of fatefulness, one belongs to the whole, one is in the whole; there is nothing which could judge, measure, compare, or sentence our being, for that would mean judging, measuring, comparing, or sentencing the whole."[54] In other words, who's to doom?

The nomad thought of *Moby-Dick* undermines both the conditions for and the aspirations of mid-nineteenth-century American national narrative. By positing an inhuman force of infinite extension over the finite American Adam, by subverting the foundations of individualism (the individual subject and its agency), and by evacuating the myth of God's mandate, of His power to reward or punish, *Moby-Dick* effectively dismantles the dominant American ideological system undergirding such notions as manifest destiny. The transvaluation of American values that emerges from these antinomic conflicts strikes, as C. L. R. James says, "at the very foundation of American civilization."[55] In producing a critique of Americanism from within—as opposed to criticizing American values from a vantage point outside of the American value system—*Moby-Dick* is able to project an alternative image of the world from that which frames mid-nineteenth-century national discourse, a discourse that maintains considerable influence today.

The nomadic trajectory of *Moby-Dick* is directly opposed to the State philosophy of the mid-nineteenth century and, moreover, to that of the twentieth century critics who wish to claim *Moby-Dick* for their own nationalist literary purposes. But, in establishing a multinational frame in the opening, unnumbered chapters and in arguing on behalf of a transvaluation of those values that lay the foundations for national narrative, *Moby-Dick* presents an image of the world that runs counter to the dominant American culture in 1851 and that anticipates the emergent postnational world system. Melville's geographical frame of reference is the world itself, typified by the open sea, "a smooth space par excellence" (as Deleuze and Guattari call it),[56] which knows no particular national allegiance, but which is also the zone of intensive international conflict, including, of course, the extra-territorial expansion of the

American imperium. "Let America add Mexico to Texas, and pile Cuba upon Canada; let the English overswarm all India, and hang out their blazing banner from the sun; two thirds of this terraqueous globe are the Nantucketer's. For the sea is his; he owns it, as Emperors own empires; other seaman have but a right of way through it." The Nantucketer, of course, is merely a sign for the workers who ride the waves, living with the sea's "unshored, harborless immensities."[57] Melville's *Moby-Dick* represents his most ambitious attempt to represent the seemingly unrepresentable totality figured in this great oceanic space. Melville's writings of the South Seas enabled him to experiment with the representational techniques and cartographic project in trying to "manhandle this Leviathan."

Chapter 6

"Spaces that before were blank": The Utopia of the Periphery

Notwithstanding its canonicity in American Studies, *Moby-Dick* is an anomaly within Melville's oeuvre, and its strangeness stands out markedly when viewed next to his previous works. With the exception of *Mardi,* in which Melville made a rather unsuccessful attempt at the literary narrative form, Melville's early works clearly lie within the personal narrative genre. If they were not entirely factual, they were at least based on personal experience, and the reader was certainly encouraged to identify the writer with the narrator or central character. In *Moby-Dick,* something changes dramatically, with respect to both form and substance (as if these could be separated decisively). The figure of Ishmael would *seem* to herald another personal narrative of the sort that Melville had written with such success, critically and financially, in *Typee* and *Omoo,* and the personal narrative form tied in easily with a national imaginary community which is undergirded in part by both individualism and national expansion. The iconic figures of national narrative—again, the American Adam, traversing a Virgin Land, on a Providential errand into the wilderness—are easily incorporated into the personal narrative form as well, enabling the voyage of a solitary narrator to figure forth a nationalist image. However, *Moby-Dick* both moves away from this iconography and produces a powerful critique of it. By transforming the personal-*cum*-national narrative form, Melville is able to project an image of a postnational world system. In Chapters 4 and 5, I focused primarily on the content of Melville's project and the ideology of national narrative; in this chapter, I would like to examine more closely the formal techniques Melville uses in his literary cartography. In exploring the global periphery and in developing a form for representing those extraterritorial spaces, Melville comes to modify both form and content in order to disclose and represent the "truth" of the world system.

Although the "Call me Ishmael" beginning of "Loomings" would seem to announce a personal narrative to follow, *Moby-Dick* quickly becomes something else entirely: a sprawlingly complex novel that attempts to encompass the entire world, "the whole circle of the sciences, and all the generations of whales, and men, and mastodons, past, present, and to come, with all the revolving panoramas of empire on earth, and throughout the universe, not excluding its suburbs."[1] Even with the tongue-in-cheek reference to the universe's suburbs, Melville simply cannot accomplish, or even attempt, such a comprehensive mapping project in the form of a mere personal narrative. The disparity between the two types of writing within the novel—a straightforward personal narrative on the one hand and an encyclopedic, supra-individual, uncategorizable and baroque work on the other—is the chief basis for the famous "two *Moby-Dicks*" theory, suggested by Charles Olson and Howard Vincent in the 1940s, and laid out explicitly by George R. Stewart in 1954. Stewart argues that an "Ur-*Moby-Dick*" tells the Ishmael story, while the later version superimposed a grander vision on the novel without substantially revising the early, onshore chapters.[2] Although this theory is dubious,[3] *Moby-Dick*'s anomalous form remains a topic of interest. Why did Melville turn away from the personal narrative form, with which he had been so familiar and successful? Why did he move away from a form based on lived experience to a form that required an almost impossible comprehensiveness? Why did he not write the story of "a whaling voyage by one Ishmael" as promised in "Loomings," and instead write a work that, as Edward Said has put it, is really about "the whole world"?[4]

The answers have to do with Melville's growing concern for what he called "the great Art of Telling the Truth."[5] Melville's obsession with the truth, discovering it and representing it, led him to abandon personal narrative in favor of a form that enabled him to produce a more comprehensive, and therefore truer, picture. In his life and in his early works, Melville found what he thought were authentic or true places far from home, in the extraterritorial contact zone of the Pacific, with its vast blank spaces on the world map. Melville's enthusiasm for the apparent novelty and unique energy of Pacific cultures enabled him to write the personal narratives that brought him early fame, but he found that the representational form was not adequate to his task. Such works failed to deliver the whole truth, which for Melville means that they failed to be true at all. In this chapter, I will argue that Melville's developed the peculiar narrative form in *Moby-Dick* as a way to represent more accurately the utopian "truth" he felt he had discovered in the South Seas. Melville's struggles with, and eventual decision to abandon, the narrative form used in his early works are both epistemic and stylistic, requiring answers to the questions, *How do I know the truth?* and *How do I represent it?* In exploring the blank spaces of his map, Melville tries to find the truth in those exotic places of the Pacific; in his quest, Melville discovers that his cartographic techniques—and the narrative form of his writing—had to change in order to present the truth as he saw it.

The figure of mapmaking is particularly apt in Melville's early South Seas writings, inasmuch as maps offer essentially fictional representations of true places, and Melville felt that he had stumbled upon true places that were "not down" on any map. In *Moby-Dick*, Melville devotes a chapter to the subject of mapping, and in it Melville introduces a figure for his entire mapping project.

The opening lines of "The Chart" depict Ahab engaged in a mapping project of his own, attempting to inscribe figures onto "spaces that before were blank," even as he himself becomes a blank space for inscription.

> Had you followed Captain Ahab down into his cabin after the squall that took place on the night succeeding that wild ratification of his purpose with his crew, you would have seen him go to a locker in the transom, and bringing out a large wrinkled roll of yellowish sea charts, spread them before him on his screwed down table. Then seating himself before it, you would have seen him intently study the various lines and shadings which there met his eye; and with slow but steady pencil trace additional lines over spaces that before were blank. At intervals, he would refer to piles of old log-books beside him, wherein were set down the seasons and places in which, on various former voyages of various ships, sperm whales had been captured or seen.
>
> While thus employed, the heavy pewter lamp suspended in chains over his head, continually rocked with the motion of the ship, and for ever threw shifting gleams and shadows of lines upon his wrinkled brow, till it almost seemed that while he himself was marking out lines and courses on the wrinkled charts, some invisible pencil was also tracing lines and courses upon the deeply marked chart of his forehead.[6]

Melville notes that, "so far as what there may be of a narrative in this book," the chapter dealing with Ahab and his charts "is as important a one as will be found" in the novel.[7] In its later part, as I discussed in the last chapter, "The Chart" undermines the sense of individuality and agency necessary for that individualism at the core of American national narrative. But in this earlier part, Melville provides a figure that might stand for the novel's overall project: filling spaces that before were blank. This image or activity appears throughout the novel,[8] but here the cartographic nature of the project is brought directly to the fore. Melville's desire to create a total world map, to fill in all the blank spaces in order to represent everything about his subject, establishes a nearly impossible task. Melville is mindful of the caveat attributed to Hakluyt and printed on the very first page of the novel: "While you take it in hand to school others, and to teach them by what name a whale-fish is to be called in our tongue, leaving out, through ignorance, the letter H, which almost alone maketh up the signification of the word, you deliver that which is not true."[9] As if fearful of delivering "that which is not true," Melville's baroque, extravagant work seems to leave nothing out, filling all the blank spaces with novel inscriptions, marking and remarking the figures in his literary cartography of the world.

The etymological caveat also reveals an anxiety underlying the global project of *Moby-Dick*. Specifically, it is the fear that truth will elude him, that despite his desired comprehensiveness, he will leave something out and the representation will be a misrepresentation. Melville's anxiety is disclosed throughout the novel, from his famous admission in "Cetology" that "This whole book is but a draught—nay, the draught of a draught" to his warning in "Of Monstrous Pictures of Whales" that no true portrait of the whale is possible, that "the great Leviathan is that one creature in the world which must remain unpainted to the last."[10] *Moby-Dick*'s caution registers the uncertainty inherent in its mapping project. In his earlier works, Melville had combined personal experience and secondary sources (travelogues and scientific works) to give "true" accounts of his adventures in the South Seas and elsewhere. Despite his "success," Melville seemed to believe that it limited his ability to find and represent the truth. In *Mardi*, Melville essayed a form—romance, rather than personal narrative—that he thought could achieve his ambitious project. Like *Moby-Dick*, *Mardi* presents a global narrative, where a literal circumnavigation opens up a series of discussions on politics, science, philosophy, and art. A key difference, *inter alia*, lies in Melville's tone. In *Mardi*, the world map Melville projects is displayed confidently, declaring its truth vociferously to the reader. After the failure of *Mardi*, which forced Melville to return to the personal narrative form in *Redburn* and *White-Jacket* ("two *jobs*," as he wrote to his father-in-law, "which I have done for the money—being forced to it as other men are to sawing wood"), a humbler Melville was much more cautious. *Moby-Dick* would retain the ambition and scope, but would express doubt about the success of its project.[11]

This "cartographic anxiety," as Derek Gregory has called it,[12] forces Melville to explore the blank spaces of the map, seeking the truth in those extraterritorial zones that he would dub "true places" and trying to find a representational form that would allow him to realize his goal of becoming a master of "the great Art of Telling the Truth." For Melville, the Pacific provided the geographical space in which the truth—a comprehensive truth that would encompass political, scientific, and the aesthetic worlds—could be revealed; his experiences in the Pacific condition both his understanding of the world and his craft of representing it. Melville's South Seas cartography registers his struggle to unite form and content. His fascination with blank spaces, true places, and spatial representation makes possible both the philosophical mediation on power in the world system and the originality of the literary form visible in *Moby-Dick*.

Blank Spaces

Joseph Conrad, in his autobiographical reminiscences, expresses the romantic joy with which, as a youth, he had penciled in spaces that before were blank on his maps. In "Geography and Some Explorers," Conrad writes of his fascination, first with polar geography, then with Africa, and he identifies the connection between the two otherwise dissimilar regions: big, white spaces.

> And it was Africa, the continent out of which the Romans used to say some new thing was always coming, that got cleared of all the dull imaginary wonders of the dark ages, which were being replaced with exciting spaces of white.

The youthful Conrad paid homage to the discoverers of the great lakes of Africa by "entering laboriously in pencil the outline of Tanganyika on my beloved old atlas, which, having been published in 1852, knew nothing, of course, of the Great Lakes. The heart of Africa was white and big."[13] For Conrad, the fascination that geography inspires is directly linked to a spirit of adventure; hence, blank spaces had a special allure:

> It was 1868, when nine years old or thereabouts, that while looking at a map of Africa of the time and putting my finger on the blank space then representing the unsolved mystery of that continent, I said to myself with absolute assurance and an amazing audacity which are no longer in my character now: "When I grow up I shall go *there*" [. . .] *there* being the region of Stanley Falls which in "68 was the blankest of blank spaces on the earth's figured surface."[14]

It is clear from Conrad's memoirs that his enthusiasm for geography, inspired above all by "blank spaces," was primarily motivated by a desire, as it were, to *un-blank* those spaces: to inscribe upon their blankness new lines and figures that come with truly *knowing* the place. "Fabulous geography," as Conrad calls it, filled the space representing unknown regions with various drawings, often of monstrous creatures, thereby masking the incomplete or downright erroneous knowledge of the area. Scientific geography—whose advent, according to Conrad, coincides with the discovery of the New World and whose spirit dominates nineteenth-century cartography—restores to maps the "honest" blankness of unknown regions. Such honest cartography, "registering the hard-won knowledge, but also in a scientific spirit recording the geographical ignorance of its time," inspired Conrad and served as a model for all knowledge acquisition.[15]

In these passages celebrating the noble adventure of geographical knowledge-production, no mention is made of the metropolitan power's direct interest in that production, but few writers knew better than Conrad of the imperialist project implicit (or explicit) in filling the spaces of the map that before were blank. In *Heart of Darkness*, Marlowe sums up succinctly: "It had ceased to be a blank space of delightful mystery—a white patch for a boy to dream gloriously over. It had become a place of darkness."[16] Of course, the cartographic inscription is a result, at one and the same time, of enhanced geographical knowledge and of direct political and economic colonization. This is the political content of the cartographic dilemma in which the "spaces of the earth's figured surface" are inscribed and re-inscribed, almost perpetually, like Ahab's laborious pouring over his charts. In Conrad, the boyish sense of wonder at blank spaces on a map is transformed into an inward-turning brooding over the darkness which had covered those spaces.[17] In *Moby-Dick*, the case is somewhat different. For the continuous revision, the tracing and retracing of courses on the map, the extensive striations of smooth space reveal, not a closed space, but a dynamic play of spatial practices—mappings—that make up Melville's literary cartography and the world it attempts to project.

As Ahab's solitary study of the charts and log-books suggests, the mapping project is epistemic, related to the acquisition and organization of knowledge.[18] Conrad's "honest maps" admitted their ignorance, but mapmaking is itself a form of knowledge, as well as an exercise of power. José Rabasa, in his *Inventing America*, makes this point when he examines the *Relaciones geográficas*, a series of questionnaires begun in 1569 and designed to obtain geographic, ethnographic, and historical information concerning Spanish America. The "epitome of Philip II's bureaucratic machinery and effort to keep control over the empire," as Rabasa puts it,

> the *Relaciones* embody the will to know indigenous geographic, political, and genealogical texts. These indigenous representations were ultimately assimilated and translated into Western conventions in the process of producing the official chronicles and maps. There is never, however, the suggestion of a view of the indigenous communities as "blank pages." On the contrary, the inquiries presuppose indigenous texts, but they also generalize the practice of writing as a form of inscribing an order of the world on a blank page.[19]

The baroque *Relaciones* functioned as a bureaucratically orchestrated gaze (in Foucault's sense of the term in *The Birth of the Clinic* or *Discipline and Punish*), extending the technologies and principles used to organize and control populations within the city or the state to the extraterritorial domains across the sea. Such surveys also extend a project in which personal or travel narratives were engaged: the collection of data about exotic locales for use in the metropolitan centers and the transcription of native experience into the language of the dominant culture.[20]

The personal narrative form is deeply bound up in an excitement or curiosity about a "new world," a world which was to be incorporated into the overall system of political and geographic thinking; put more simply, narratives of the new world placed these regions firmly on the map. There are obvious family resemblances between American personal narratives of the mid-nineteenth century and the numerous sixteenth- and seventeenth-century reports from the "New World," which were among the most popular works of literature at the time. As the European nation-state was consolidating its strength in such places as Spain, the Netherlands, England, and France, it was also radiating its power outwards, as if by centrifugal force, to the "ends" of the earth and sea. This extraterritorial extension is not, however, secondary to the state's formation; rather, it is a key part of that formation. As Werner Sombart put it, "the modern state emerged from the silver mines of Mexico and the gold mines of Brazil."[21] Travelogue and utopian literature, the two forms which most directly address the new world phenomena and which are generically close relatives of personal narratives, play a significant part in the integration of the extraterritorial into the imaginary space of the still-in-formation nation-state, and thus may find themselves incorporated into national narrative as well.[22] It is not surprising, then, that the personal narrative form achieves such popularity in the United States during the 1840s. Of course, many personal narratives antedate this moment and the territorial and extraterritorial expansion of the American national space had been well under way for some time, but this decade witnesses a remarkable acceleration in the movement of the nation's own centrifugal forces. This acceleration is visible in the nation-state's direct appropriation of territory (for instance, after the Mexican-American War), in its ideology of manifest destiny, in its major demographic shifts that include a westering population (still largely only Midwestern, but typified by the California Gold Rush), and in its mercantile and military exploits outside of North America, such as those chronicled in Melville's early works. It is from within this national and international context that personal narratives emerged and flourished.

A key aspect of personal narrative is its profoundly temporal dimension, albeit a dimension that refigures time, projecting temporality onto a spatial array. That is, the new world space to be integrated into the world of nations is understood as somehow frozen in the past, temporally displaced from the ongoing rush of history. Personal narratives functioned in such a way as to bring such places into the modern world. The absorption of these places into the political geography of a world system involved, at least rhetorically, a sense of accelerating the historical development of these regions. This is a key aspect of the *mission civilisatrice*: more than bringing salvation to heathens and developing markets or extracting natural resources (yet underlying all of these practices), the *idea* is that the geographically distant is also historically behind; it is the duty of the civilized to bring these others up to date.[23] Personal narratives may mobilize this idea in a number of ways, and the narrative form is itself a way of accomplishing such colonization.

In Melville's personal narratives, the descriptive detail is typically interspersed with ironic asides in which the events, customs, or objects of the "exotic" sphere are correlated to those of the mainstream American culture. And Melville, of course, more often than not (at least in the South Pacific narratives) uses the foreign or the exotic to criticize mainstream America or Europe. One might call Melville's method *contrapuntal*, after Edward Said's use of this musicological term in *Culture and Imperialism*.[24] The comparative analysis of cultures typifies Melville's approach throughout much of his career, and Melville frequently attempts to dismantle the opposition between "civilized" and "savage" that so characterizes narratives of the time. In *Typee* and *Omoo*, Melville almost constantly bounces back and forth from citations of conventional American or British manners or customs to descriptions of those he finds in the South Seas. This comparison by itself is not surprising, since the personal narrative form often functions as a means of bringing modern, "civilized" readers into contact with the remotely foreign culture; this contact may be accomplished through analogy, by gesturing towards what the readers already know in order to show them the significant differences of that which is unknown or unfamiliar. But Melville does not so much accentuate the differential identities between the two groups; instead he often chooses to show that the identity is itself hybrid. That is, there is no truly *authentic* culture in Melville's work, notwithstanding numerous references to an idyllic scene or to one culture's adulteration of another. Certainly Melville decries the mistreatment of the Pacific peoples by European missionaries, merchants, and militaries,[25] but it is misleading to think of these abuses as the corruption of an otherwise "pure" people or culture. In fact, such thinking is more in line with that of the evangelicals who worry that the "pure" of heart would be contaminated by books like *Typee* (see below). Whether dealing with "noble savages," like Mehevi or Kory-Kory, or such celebrated American heroes as John Paul Jones, Melville always affirms the inter-connectedness of civilization and savagery, and in so doing he gently mocks the ruling ethos of the *mission civilisatrice*.[26] It is this sort of thinking that aroused the ire of Melville's most ardent critics who charged him with idealizing primitive cultures or "savagism."[27]

The contrapuntal approach highlights the degree to which these extraterritorial spaces were not really blank. As Melville takes pains to point out, the customs and mores of the Marquesans were as historically valid as those of the Americans. If one could not find in these supposedly blank regions a clean slate, a fresh and unspoiled Eden, one could nevertheless seek a kind of authentic experience, something more real or more true than the experiences of the dominant metropolitan culture—a theme sounded loudly in Henry David Thoreau's personal narratives, for example. This concern for authentic truth is a defining characteristic of the personal narrative form, and this contributed in no small way to the form's popularity. As narratives of displacement from mainstream American life, providing detailed descriptions of worlds foreign to that of their readers, books like Melville's *Typee*, Richard Henry Dana's *Two Years Before the Mast*, or Frederick Douglass's *Narrative* offered vistas into experiences and places that were somehow more real, more *true* by virtue of the narrator's own authority, than the world to which the readers were accustomed.

True Places

"Queequeg was a native of Kokovoko, an island far away to the West and South. It is not down on any map; true places never are."[28] Queequeg's biography offers one of many occasions in *Moby-Dick* for Melville to reflect on the inadequacy, if not downright impossibility, of representation. Indeed, this sense that the true cannot be adequately represented, here figured precisely in terms of a mapping project, is a central feature of Melville's literary cartography. In his early works, Melville grappled with this problem, and the dynamic tension between a faithful recounting of facts and the need for speculative solutions to the crisis of representation animates these personal narratives. In a quite obvious sense, the *place* represented on the map is not *true*; there can be no perfectly mimetic image of the *topos* (Borges's fabled imperial cartographers notwithstanding). In *Redburn*, as I discuss in the next chapter, Melville presents a comical example, as young Redburn "grows intolerably flat and stupid over some outlandish old guidebooks" while noticing that his map of Liverpool "bore not the slightest resemblance" to "the identical place itself."[29] But maps, however "untrue" in the abstract, must also be drawn. The question, then, is how to depict "true places."

It is no accident that the geographical *locus classicus* of this cartographic anxiety is found, in Melville's works, in the South Seas. This "contact zone" (to use Mary Louise Pratt's term)[30] presents greater than usual problems for the literary cartographer. The expansive region is historically misrecognized, and therefore misrepresented, in part because of the "uncenteredness" of the place, which, without clear representational focus, appears a kind of *no-place*. The utopian character of Melville's early personal narratives owes much to the utopian character of his subject—that is, the terrain to be mapped. In his lecture on "The South Seas," Melville notes that the very name of the ocean derives from a fundamental mistake owing to its discoverer's flawed perspective. As Melville states flatly,

> "South Seas" is simply an equivalent term for "Pacific Ocean" [. . .] But since these famous waters lie on both side of the Equator and wash the far northern shores of Kamchatka as well as the far southern one of Tierra del Fuego, how did they ever come to be christened with such a misnomer as "*South* Seas"? The way it happened was thus: The Isthmus of Darien runs not very far from east to west; if you stand upon its further shore the ocean will appear *south* of you, and were you ignorant of the general direction of the coastline you would infer that it rolled away wholly towards that quarter. Now Balboa, the first white man who laid eyes upon these waters, stood in just this position; drew just this inference and bestowed its name accordingly.[31]

This passage points to a number of problems inherent to personal narrative, problems that would eventually lead Melville to abandon the form in favor of literary narrative. Personal narrative necessarily requires the individual subject to guarantee its authenticity, and the subject necessarily has an incomplete knowledge (in this case, of geography); moreover, this already incomplete knowledge is further limited by one's particular perspective: as in the case of Balboa's misprision, personal narrative requires a kind of eye-witness inspection that cannot really comprehend the geographic totality. Hence the very *seeing*, which had been the most authoritative sense (as with eyewitness testimony), becomes an impediment to the "true" representation. This reliance upon individual perspective, which had served to lend authority to the narrative, ends up distorting the truth. As Melville goes on to point out, when Magellan redubbed the ocean "Pacific," it was merely because, having passed through the dangerous straits which bear his name, Magellan had the good fortune to encounter pleasant weather on the other side. "But the names were owing to first impressions, and the [. . .] great sea hence will forever be called Pacific, even by the sailor destined to perish in one of its terrible typhoons."[32] Melville's remarks suggest that to capture the authentic experience, you must go there yourself (that is, after all, what made Melville an authority on the South Seas in the first place), but personal experience inevitably produces erroneous or misleading information. One's knowledge must remain incomplete, and therefore untrustworthy.

Knowledge, or rather a certain kind of knowledge, seems to be the real subject of personal narratives. The word *narrative* derives from the Latin verb *narro* ("to make known," as well as "to tell or narrate"), itself closely related to the adjective *gnarus*, or "knowing," which in turn derives from the Greek *gnosis* ("knowledge"). In its original usage, *narrative* referred more to testimony than to storytelling. Significantly, in *The Emergence of American Literary Narrative*, Jonathan Arac points out this etymological background only in the chapter devoted to *personal* narratives. Personal narratives are expected to be *true:* so stringent is this requirement that, famously, the manuscript for *Typee* was rejected by Harper and Brothers on the grounds that "it was impossible that it could be true and *therefore* was without real value." This from the publishers of Poe's *Narrative of Arthur Gordon Pym*![33] As is well known, the initial controversy surrounding *Typee* was over its veracity. Melville's London publisher demanded documentation; many reviewers, mostly favorable in their estimation of the narrative, nevertheless concluded that *Typee* cannot have been a true story; and the hubbub over the *Typee's* "truth" only settled down after Richard Tobias Greene, "Toby" of the narrative, emerged in Buffalo and vouched for Melville's account.

Why such concern over the truthfulness of this book? A standard, if partial, answer is that fiction or "romance" had but little value for the readership of the day, a plausible explanation but one whose force is undermined by the generally enthusiastic response to the novels of a Cooper or a Dickens, not to mention the immense popularity of works by those Nathaniel Hawthorne dubbed the "damned mob of scribbling women." The controversy over *Typee*'s veracity is related to a secondary, but no less impassioned, controversy. Greene's credibility-restoring letter to the editor of the *Buffalo Commercial Advertiser* (published 1 July 1846) points to this. Referring to a review of *Typee* in the *New York Evangelist* (the one which had accused Melville of "savagism"), he writes:

> The *Evangelist* speaks rather disparagingly of the book as being too romantic to be true, and as being too severe on the missionaries. But to my object: I am the true and veritable "Toby," yet living, and I am happy to testify to the entire accuracy of the work.[34]

The question of the narrative's truthfulness is thus complicated with the direct critique in the text of the baleful effects of the evangelical civilizing mission. If *Typee* be mere romance, then—so the anonymous *Evangelist* reviewer suggests—its critical "slurs and flings against missionaries and civilization"[35] should not be taken seriously. *Typee*'s critique of missionaries cuts also into the fabric of American national ideology: the South Seas missionaries are merely continuing that divinely ordained mission (the errand into the wilderness) of the American people.

Of course, *Typee* and Melville's other personal narratives *are* largely fictional. Though they are loosely based on personal, lived experience, Melville by no means attempts to set forth "just the facts." On the contrary, it is clear that far more fiction than fact pervades these narratives.[36] Yet, unlike Hawthorne, who in the prefaces to his romances peremptorily defends his work against those who would read them as true, Melville faces the readers and critics as an ardent truth-teller, even when he exaggerates the truth quite a bit. Hawthorne's prefaces have a rather defensive and ironic tone, warning the reader not to connect the story with any real world experience. For example, in the preface to The Blithedale Romance, a novel which comes as close to any of Hawthorne's to being a personal narrative, since it is loosely based his experiences at the Brook Farm commune, Hawthorne writes:

> In short, his [i.e. the author's] present concern with the socialist community is merely to establish a theatre, a little removed from the highway of ordinary travel, where the creatures of his brain may play their phantasmagorical antics, without exposing them to too close a comparison with the actual event of real lives. In old countries, with which fiction has long been conversant, a certain conventional privilege seems to be awarded to the romancer; his work is not put side by side with nature; and he is allowed license with regard to every-day probability, in view of the improved effects which he is bound to produce thereby. Among ourselves, on the contrary, there is as yet no such Faery Land, so like the real world, that, in a suitable remoteness, one cannot well tell the difference, but with an atmosphere of

> strange enchantment, beheld through which the inhabitants have a propriety of their own. This atmosphere is what the American romancer needs. In its absence, the beings of imagination are compelled to show themselves in the same category as actually living mortals; a necessity that generally renders the paint and pasteboard of their composition but too painfully discernible.[37]

This passage, in a manner similar to that of Hawthorne's other disclaimers,[38] delineates a crucial distinction between what Arac calls *literary narrative* and other narrative forms. Hawthorne seems to imagine two kinds of unsympathetic reader: one who will find fault in the work for its not being realistic enough, and who will thus criticize the plot for its improbability and the characters for their "paint and pasteboard" composition; and another who will charge that the book aims at fact (and fails to reach the mark) or deliberately misrepresents real people and places, a form of perjury or a "literary crime." These are the same readers with whom Melville had to deal, and the initial criticism of *Typee* reflects this: either the story is too improbable to be true (as Harper and Brothers and many reviewers thought), or it is an attempt to slander real people (as the *Evangelist* review suggested).

T. Walter Herbert reads Melville's *Typee* as a romance rather than as a strictly personal narrative; Herbert argues that the romantic impulse is by nature geared towards undermining the established values of, in this case, religious authority, though it is but a short jump from the religious to the juridico-political authorities of an increasingly dominant (and largely secular) nationalism. In *Marquesan Encounters*, Herbert maintains that *Typee*'s "romantic" character represented the greatest danger to the dominant social values, a greater danger perhaps than even the direct attack on those values.

> At stake in the *Evangelist*'s suspicion that *Typee* is "sheer romance" is the effort of orthodox believers to keep their systems of interpreting moral reality firmly in place. The imaginative freedom of romance, which permits alternative ways of viewing experience to be indulged, threatens to undermine the absolute authority that the orthodox ascribed to their moral vision. The deviant forms of thought and behavior represented by romances may describe things that really happen among men, but they should only be acknowledged in the act of arresting or condemning them: they "should be known only to the police."[39]

This view of romance does not draw a distinct line between fact and fiction or between the imaginary and the real, as Hawthorne's disclaimers try to do, but rather, romance is here viewed as another, competing form of knowledge-production and dissemination, a form which cannot be trusted when in the wrong hands.

This digression on romance leads me to a paradox: on the one hand, the personal narrative form can be subversive because it is true, an authoritative setting forth of facts, facts that in turn might fly in the face of reigning myths developed in order to hold together a certain vision of society;[40] on the other hand, Melville's personal narratives are also romantic, or at least they may function as romances, and hence they could be viewed as presenting false or misleading information to a reading public whose imagination is stirred by the "pretty lies" of fiction. Whether one is faithfully recounting matters of fact or creating a "theatre" where imaginary characters "may play their phantasmagorical antics," you're damned if you do and damned if you don't.

Melville was to find personal narrative's injunction to tell the truth onerous, even though few writers of the mid-nineteenth century could be said to be more concerned with "truth-telling" than he. As a young man, Melville fancied that he had found true places in the South Seas. For Melville, *truth* is always related to *space*, and consequently, the ascertaining of truth, and truth-telling or narrative, are imagined as geographic or cartographic enterprises. In "Hawthorne and His Mosses," a panegyric to the "great Art of Telling the Truth," Melville writes:

> You must have plenty of sea-room to tell the Truth in; especially, when it seems to have an aspect of newness, as America did in 1492, though it was just as old, and perhaps older than Asia, only those sagacious philosophers, the common sailors, had never seen it before; swearing it was all water and moonshine there.[41]

If the Pacific contact zone, containing a great deal of water and moonshine, offered Melville a *place* for telling the Truth, literary narrative would present the *form* for telling it. Notwithstanding the clear success he had enjoyed with *Typee* and *Omoo*, Melville turned away from the personal narrative form with *Mardi*, an audaciously ambitious romance, set at first in the South Seas but quickly launching into a global circumnavigation. By its attempt, Melville was indicating that the space of literary narrative was far larger than local or national space; in attempt to comprehend this larger space, Melville produced what Franco Moretti calls a "world text" with the world system itself as its geographic frame of reference.[42] Such a geography would have to go well beyond a narrator's personal experience. The scope would have to be much greater than that which could be seen with the naked eye.

Projections

The adventurous and scientific spirit that made Conrad pour over the big, white spaces of his "honest" maps, inscribing true places into the spaces that before were blank, is not unlike the spirit animating the literary writer. Both geographer and writer work with imaginary and real materials, raise questions or scope and scale, decide how much detail to include, and so on. Peter Turchi has suggested that all writing is essentially a form of mapmaking, and the writer is thus a cartographer.[43] In Melville's case, the personal narrative form proved inadequate to the task he had set for himself, which involves a comprehensive, global representation. For Melville, the truer map would come from literary narrative.

The personal narrative form, depending as it does on the eyewitness account of a single narrative voice, is really more of an itinerary than a map, strictly speaking. This distinction is made by Jameson in his discussion of cognitive mapping, specifically with respect to the practices set forth by Kevin Lynch in *The Image of the City*. Lynch had described the sense of alienation felt by individuals in modern cities, particularly those cities that had no clearly recognizable landmarks by which to orient oneself. The ability to map one's environment cognitively would seem to solve that problem. As Jameson puts it, overcoming the alienation of the traditional city, then, involves "the practical reconquest of a sense of place and the construction or reconstruction of an articulated ensemble which can be retained in memory and which the individual subject can map and remap along the moments of mobile, alternative trajectories."[44] But Jameson notes that, in Lynch's formulation, this activity is really "precartographic," resulting in an itinerary rather than a map. Itineraries are "diagrams organized around the still subject-centered or existential journey of the traveler, along which various significant key features are marked." In contrast, maps employing artistic and technical practices (for example, using such tools as the compass and the sextant) introduce a relationship with a nonsubjective totality: "cognitive mapping in the broader sense comes to require the coordination of existential data (the empirical position of the subject) with unlived, abstract conceptions of the geographic totality." Finally, with the developments of the globe and the Mercator projection, cartography involves more directly the question of representation itself; that is, mapmaking abandons its "naively mimetic" aspirations in favor of more complicated meditations on representational form. "At this point it becomes clear that there can be no true maps (at the same point it also becomes clear that there can be scientific progress, or better still, a dialectical advance, in the various historical moments of mapmaking)."

Jameson's brief sketch of the history of cartography highlights an issue that Melville was grappling with in his own work: how to connect the individual experience of the world with the larger social and geographic totality. In *Moby-Dick*, Melville clearly endeavors to go beyond the limits of his earlier personal narratives. The personal narrative that appears to begin with the words "Call me Ishmael" is quickly revealed to be a ruse.[45] Such a narrative is overwhelmed by the world system projected throughout the novel, where boundaries are forever crossed, frames perforated, and explanatory categories rendered inadequate. In attempting to create a comprehensive work, one that could encompass the entire world, Melville had to alter the form and the content of his writing. He had to develop a baroque literary narrative that veered from the truth-telling rules of personal narrative, while projecting a world system beyond any one individual's ken. This would involve excess, exaggeration, and speculation. Melville decided that, to get at the Truth, one must abandon one's allegiance to the straightforward setting forth of facts.

The Mercator projection offers an interesting analogy. Mercator's projection was used to produce his influential world map in 1569. Famously (or infamously), the Mercator projection distorted the geographic spaces it purported to represent; objects appeared larger the further they were from the equator, thus grotesquely aggrandizing the land masses near the North Pole. (For example, the Mercator projection is responsible for those maps depicting Greenland as roughly the size of South America, even though it is actually about one-sixth of the size of the southern continent.) Although such a map may have benefited Eurocentrism in the long run, Mercator's own goal was somewhat more prosaic.[46] The projection was developed in order to solve the technical and practical problem of transferring curved space to a flat chart. This is not merely a mathematical and artistic issue, but essential one for confident navigation over long distances. With maps using the Mercator projection, sailors could set a course using straight lines, thus establishing truer courses. Obviously those navigators would realize that the map was not "true," in the sense of mimetically accurate in its depiction of the places figured on its surface. The fictional map, however, provided the better view of the world.

Melville's literary cartography, like Mercator's, requires the intervention of fancy, of imaginative or speculative writing that looked rather different from the work which preceded it. It is almost a cliché to say that fiction presents a greater truth than fact, but Melville's development of a literary narrative form enabled him to attempt the project that he felt necessary for accomplishing the great Art of Telling the Truth. Like Hawthorne, Melville needed to create a world apart; yet, unlike Hawthorne, who seemed content to present his tragedies and farces at an imaginary roadside theater, Melville was to remain assiduously in the world, insisting on the necessity of a global, comprehensive mapping project, while agonizing over the inevitability of his failure. In a letter to Hawthorne, written after the publication of *Moby-Dick*, Melville sighs, "As long as we have anything more to do, we have done nothing."[47] Melville peers into the blank spaces of his map, like the white pages of the manuscript not yet written and like the inscrutable white skin of Moby Dick himself, etches into them lines in spaces that before were blank, and emerges with a form suitable for representing the unrepresentable. There can be no true maps. But through the patient, meticulous writing and rewriting, there can be infinite mappings.

Melville's literary education in the South Seas makes possible his renunciation of the personal narrative form and his embrace of a narrative from that would enable his global literary cartographic project in *Moby-Dick*. His critique of personal narrative and the failed representational project it entails did not, however, end there. After a foray in to the tombs of the American unconscious in *Pierre* and the heart of American capitalism in his "story of Wall Street," Melville returns to the South Seas with a vengeance in what may be his most powerful post-*Moby-Dick* work, "Benito Cereno," in which he decidedly puts to rest the truth-telling pretensions of personal and national narrative. Where the latter forms claimed to reveal detailed pictures, drawing distinct black lines upon the hitherto blank spaces of the map, shedding daylight upon the dark places of the world, and establishing the utopia of the periphery, Melville's enshrouds these in the grey chaos that better approximates the experience of the emerging world system. In "Benito Cereno," which is based on an actual personal narrative by the historical Amasa Delano, Melville upends the personal narrative form, showing how its pretensions to clarity only lead to further confusion. In slowly, patiently, carefully, and meticulously revealing the "truth" of Babo's and Atufal's masterfully executed deception, Melville fashions a powerful literary narrative in clear opposition to the personal narrative form. That he does so while positing a typical (perhaps archetypical) American as the befuddled and naïve captain, and further draping the overthrown Spaniards in the mantle of the national and imperial (with no lesser a symbol of America's national origins than Christopher Columbus himself as the literal and figurative figurehead), Melville underscores the critique of national narrative implicit in the critique of the personal narrative form. In Babo, defiant even in death as his head ("that hive of subtlety") faces down the ailing Don Benito, Melville figures forth a force—not unlike what Michael Hardt and Antonio Negri term *the multitude*—to challenge the project of imperial nationalism.[48]

Before returning to the Pacific contact zone in *Moby-Dick*, "Benito Cereno," and "The Encantadas," Melville had to contend with another distinct spatial organization. Perhaps even more profound and transformative than the exotic, extraterritorial and seemingly utopian spaces incorporated into the conceptual geography of the nation-state, the burgeoning urban space of the city confounded both the expectations of lived experience and the writer's ability to represent it. In mapping the city, Melville had to refine his literary cartography.

Chapter 7

A Prosy Stroll: Overview and the Urban Itinerary

In Melville's earlier explorations of the wide-open spaces of the Pacific, the personal narrative itinerary gradually revealed itself to be inadequate. Melville sought different techniques to represent a world system that could not simply be read by an individual subject, but must be projected in an effort to give form to, or make sense of, its baroque complexity. The extraterritorial zone of the South Seas offered a productive *topos* for Melville's literary cartography, but another representative spatiopolitical zone was also emerging around the same time, an urban space that increasingly came to define the contours of the personal, national, and global geography as well. In his grappling with the ambiguities of literary narrative in the midst of an emerging world system, Melville also had to wrestle with the ambiguities of the urban.

Prior to *Moby-Dick*, Melville had already begun the process. Although the settings of his early works are not really urban—only in *Redburn* does a character explore a major city—the atmosphere becomes more and more urban, as in Melville's description of a naval warship as a "city afloat" in *White-Jacket*. Moreover, the sequence of Melville's five books written prior to 1851 exhibit a pattern of increasing totalization which was to be realized only in *Moby-Dick*, and which was not to be quite so fully realized again (although it is visible in the literary narratives of the 1850s, including "Bartleby, the Scrivener" and "Benito Cereno"). This pattern can be seen in the changing role of the writer vis-à-vis his subjects, in the scope of the literary projects, and in the function of literature itself. In particular, as discussed in the last chapter, Melville abandons the very popular and successful personal narrative form in favor of literary narrative, and for Melville this means more than merely turning away from personal experience in the "real" world and toward a more "imaginary" one. The move to the literary narrative form enables Melville to project a literary cartography on a global scale, freed from the perceived constraints of national and personal narratives, liberated from their inherent links to images of an American Adam, a Virgin Land, and an errand into the wilderness—liberated, that is, from the American State philosophy of his day. Perhaps the greatest challenge to the *mythos* of American national narrative was the rise of densely populated cities in the mid-nineteenth century, where the notion of Virgin Land and errands into the wilderness must either be abandoned entirely or become even more metaphorical. In addition to defying certain ideological conventions, the newly aggregated or concentrated urban spaces presented new challenges to literary cartography. Novel urban experiences needed new forms of representation.

Melville's movement from personal to literary narrative in the 1840s is not one of continuous development. After the success of *Typee* and *Omoo*, Melville's first attempt at literary narrative, *Mardi*, was a disastrous failure (both in the marketplace and in Melville's own judgment).[1] With *Redburn* and *White-Jacket*, Melville returned to the personal narrative form, and even after the ambitious literary narratives, *Moby-Dick* and *Pierre*, Melville once more produces a personal narrative of a sort in *Israel Potter* (1855), along with his other magazine writing. But even within the personal narrative form, as seen in the pre-*Moby-Dick* works, Melville was already extending the scope of his literary cartography. As noted in the previous part, this expansive movement, in which the range of experience represented goes from the displaced individual's observations of a particular place to a grander exposition that surveys an entire domain (the "world")—the shift from personal to literary narrative—is analogous to a shift in Melville's literary cartography from itinerary to map. In Melville's early works, this movement is also related to a growing urbanism, or urban atmosphere, in the narratives. The ambiguities of the urban create new experiences and new effects, which in turn require new ways of representing them. Melville's formal considerations are also affected by the content, as spaces depicted produce entirely different effects on the subjects inhabiting them.

A Need for Overview

The plots of Melville's personal narratives follow a basic "descent-and-return" pattern characteristic of the form as it reached the height of its popularity in the 1840s. This formula, according to which a narrator of a relatively high social class (if not necessarily a high economic class) descends into a lower strata of society or a subordinate domain outside of the society (e.g. into poverty or "savagery") only to return to his former position to relate what he or she had observed, is typical of personal narratives as a genre, but it is also familiar to readers of fantasy or adventure stories, and the formula contributes to the character of the *Bildungsroman* as well. A chief distinction between American personal narratives of the 1840s and the *Bildungsroman*, a literary narrative, is that the former tends to lack that interiority, the deeply psychological development of the protagonist or narrator.[2] In modern novels, as both Georg Lukács's and Mikhail Bakhtin's theories of the novel assert, the "action" is largely identical to the development or transformation of the principal character or hero, so that the psychological transformation of the hero is what the story is really all about. However, in personal narratives the protagonist or narrator merely reports on events, places, and people without necessarily undergoing any transformation in any deeply personal way himself. To be sure, the movement from innocence to experience that typifies the *Bildungsroman* and that lends its authority to literary narrative may be found also in personal narratives, but compared with the literary or novelistic heroes, the personal narrative's reporter does not alter much, and his own psychological development rarely factors into the narrative at all. The descent-and-return formula, then, is less likely to produce grand epiphanies—Dante the Pilgrim's journey is perhaps the ultimate example—than it is to appear to be a moment of "slumming," a bit of an odd holiday before returning to the world of normal, quotidian experience.[3]

Melville wearied of the personal narrative form. Notwithstanding his success with *Typee* and *Omoo*, Melville wanted to devise strategies for comprehending a greater field than the personal narrative form would seem to allow. In *Redburn* and *White-Jacket*, Melville changes the way the narratives unfold, though still working within personal narrative. In *Redburn*, for instance, the first-person narration is maintained, but now the third-person voice of the chapter titles indicate a rupture between the book's subject and the book's author, even if *Redburn* is perhaps Melville's most autobiographical book. Thus, *Redburn's* Chapter 10, "He is Very Much Frightened; the Sailors Abuse Him; and He Becomes Miserable and Forlorn," could end with the line, "And there I sat, till at length the dull beating against the ship's bows, and the silence around, soothed me down, and I fell asleep as I sat."[4] In *Typee* and *Omoo*, by contrast, the narration and the chapter headings remained in the same voice. By this move, Melville attempts to gain an overview of his subject. Wellingborough Redburn will tell the story, but, while he is doing it, we the readers will have a slightly better vantage, looking over his shoulder as it were, and seeing a bit more than the narrator's personal vision can. With *White-Jacket*, Melville's dissatisfaction with the limitations of the form is clearly visible, as in the book he attempts to produce an image of the world ("the world in a man-of-war," as the subtitle states), and, in order to do this, he tries to develop a much greater field of vision, an overview capable of taking in at once an organized totality, even if the narrator never leaves the ship. But this was not the first time Melville made this attempt.

The writer of *Mardi* was, in many ways, not the same as the writer of *Typee* and *Omoo*; at least, he did not think that he was. In a letter accompanying his submission of the manuscript to his publisher, John Murray, Melville writes: "Unless you should deem it very desirable do not put me down on the title page as "the author of Typee and Omoo." I wish to separate "*Mardi*" as much as possible from those books."[5] In an earlier letter to Murray, he admits that he had planned to write another travel narrative, a continuation of *Omoo*, but "proceeding in my narrative of *facts* I began to feel an incurable distaste for the same; & a longing to plume my powers for a flight, & felt irked, cramped & fettered by plodding along with dull common places."[6] This admission is quite telling, as Melville himself alludes to the distinction between personal and literary narrative. It is not merely the difference between straightforwardly recounting facts and mobilizing more imaginative materials, since Melville had always played fast and loose with the facts anyhow. Melville's metaphors in this letter point to a more crucial distinction: the power of flight versus mere "plodding along," freely soaring above versus laborious walking below. Flight enables the writer to rise above the "dull common places" of "the same," and to survey a broader terrain than had been possible in Melville's earlier efforts. Hence, *Mardi*, as its author asserts, is not merely a fictionalized version of *Typee* or *Omoo*, but it "is made of different stuff altogether."[7]

The distinction between "plodding" and "flight" is also typical of the urban experience Melville will explore. A crucial part of the representational crisis in urban cartography—that is, with *cognitive mapping* in the city, in Kevin Lynch's or Fredric Jameson's sense—is the bewildering experience of walking on the dense, crowded streets. Such bewilderment is compounded if there is no sense of overview with which to orient oneself. Is this not what a map purports to be? A bird's-eye view of the space one must try to navigate? Just as Melville found the itinerary of South Seas personal narratives to be inadequate to the task of mapping a totality of experience (i.e. the only way "truth" could be *revealed*), so too is he unable to map urban space properly while merely walking around on the ground. As I will argue below, however, Melville's attempt to represent the cityscape using the same techniques he had developed in other literary narratives founders on the shoals of the ambiguous urban space.

Melville metaphor of flight is also well suited to Melville's purposes in *Mardi*, for his first attempt at real romance would attempt a global map made possible only by establishing a grand overview of the entire world. *Mardi*, along with Edgar Allan Poe's *Eureka*, are among the first literary narratives to appear in the United States, and each of these works takes "the world" as its subject and as its representative space. *Eureka*, a work dedicated to the Baron von Humboldt (whose scientific research and travels in "New Spain" is emblematic of the centrifugal forces of European statecraft at the turn of the nineteenth century, opening up a new world), is self-consciously a world text,[8] taking the whole universe (perhaps, not excluding its suburbs) as its subject. *Mardi* is also a world text, and Melville's "Polynesian romance" turns out to be nothing less than a global mapping project, no longer limited to the South Seas. So, for example, early in the novel, the reader learns that the very language spoken (that of an "aboriginal tar") is not national, but cosmopolitan: "You sink your clan; down goes your nation; you speak a world's language."[9] A crucial turning point in the book—specifically, the point at which *Mardi* begins to turn away from the formal structure of personal narrative into something else—is marked by a chapter titled "World Ho!" After this point, the "plot" is overrun by literary, philosophical, and political speculations, and the voyage becomes a journey around the world. The circumnavigation takes in such contemporary, though world historical, events as the Paris Revolution of 1848; the indefatigable speakers aboard Melville's ship raise political questions over Irish nationalism, slavery, and monarchical versus republican forms of government. This wild, often disorganized novel is very much of the world and still participating, if only through seemingly infinite conversation, in it.[10]

A defining characteristic of mid-nineteenth-century literary narrative is its attempt to put the real world at bay in favor of an imaginative world of the author's own creation, as Hawthorne's prefaces insist. "The result of successful imaginative activity is the literary creation of a world that stands free from the world of everyday life as an independent whole."[11] Certainly the imaginative creation of a "new world" is one of Melville's goals in *Mardi*, but his rambling adventure touches on so many places and topics, barely disguised, which were vital to political and cultural discourse in the late 1840s that Melville can hardly be said to have turned away from quotidian reality. Rather, *Mardi* represents Melville's first sustained effort at projecting an image of the world system as he imagined it, and it clearly paved the way for the literary cartography of *Moby-Dick*. *Mardi* is a literary narrative, an attempt to create a "new world" and to organize heterogeneous elements into an integrated whole, or a totality. Still writing very much under the sign of personal narrative, however, the representation of this totality gets away from the author frequently in the text. Melville's extravagances are present in nearly all of his writings, but in *Mardi*, they consume the text, constantly hindering the achievement of the desired totalization. As a form-giving-form, *Mardi* failed to accomplish the overview required to make sense of the chaotic ensemble of fragments it encounters and creates. Yet Melville, perhaps aware of this failure, nevertheless realizes that this literary narrative is what he wants to produce, and it is of an entirely different order than the personal narratives with which his name had been made.[12]

The distinction made by Melville between *Mardi* and his two preceding works points to that between a map and a mere itinerary. In the last chapter, I referred to Jameson's distinction between the pre-cartographic itineraries, "diagrams organized around a still subject-centered or existential journey of the traveler," and proper maps, which "require the coordination of existential data (the empirical position of the subject) with unlived, abstract conceptions of the geographic totality."[13] The personal narrative form is analogous to the itinerary, insofar as its authority lies in the person of the individual subject, the narrator, whose perspective is necessarily limited, as Melville himself indicates. Melville's transition to literary narrative may then be read as an effort to coordinate the individual's experience with a geographic and social totality to which no single individual has access. All of Melville's works are in some sense cartographic, but it is with his literary narratives that the literary cartography approaches its desired comprehensiveness.

Mardi teaches Melville that, in order to produce the sort of work he wants, he needs to come up with new techniques. He cannot simply substitute an imaginary world of his own creation that will resemble the "real world." Rather, he must *project* a world. As with the Mercator projection, Melville's cartographic project must transform reality in the effort to represent it. This means, in part, giving up the pretension of being able to perfectly, mimetically represent that world, turning away from a "narrative of facts," and risking presenting a "monstrous" picture. Indeed, one could argue that part of the failure of *Mardi* lies in the rigidity of its geopolitical allegory; its imaginative world too faithfully mirrors the image of the real world, albeit with the place-names changed. The literary narrative form could enable truer maps.

Of course, between *Mardi* and *Moby-Dick*, Melville produces two more personal narratives, two more itineraries—*Redburn* and *White-Jacket*—written in a matter of months in 1849. Melville's taste for "different stuff altogether" is apparent in both his haste in writing these novels and his disdain for them and their potential success. In a famous letter to his father-in-law, Lemuel Shaw, Melville writes:

> For Redburn I anticipate no particular reception of any kind. It may be deemed a book of tolerable entertainment;—& may be accounted dull.—As for the other book, it will be sure to be attacked in some quarters. But no reputation that is gratifying to me, can possibly be achieved by either of these books. They are two jobs, which I have done for the money—being forced to it, as other men are to sawing wood [. . .] Being books, then, written in this way, my only desire for their "success" (as it is called) springs from my pocket, & not from my heart. So far as I am individually concerned, & independent of my pocket, it is my earnest desire to write those sort of books which are said to "fail."[14]

These are clearly not the books Melville would have written had not the failure of *Mardi* and the need for cash—he was now married, with children—influenced his plans. It is not surprising, therefore, that *poverty* is a key theme of both works. With *Redburn*, he promised "no metaphysics, no conic-sections, nothing but cakes & ale."[15] But the experience of *Mardi* had left the writer unable to return to the jaunty narratives of personal experience that had won him fame. *Redburn* and *White-Jacket* are darker than their predecessors, and they introduce a new element, a new representative space, into Melville's literary cartography.

Redburn: His First Voyage is as much a tale of a young man's experiences with urban life as it is a story of his first sea voyage. Redburn's journey begins as a trip from upstate New York to New York City, and the account of his time in the thriving industrial seaport of Liverpool takes up nearly a third of the book. *White-Jacket,* by contrast, takes place exclusively on board an American man-of-war; even when the ship is in port, the narrative remains strictly shipboard. Nevertheless few of Melville's books are as rife with urban imagery as *White-Jacket,* for the "man-of-war is a city afloat, with long avenues set out with guns instead of trees, and numerous shady lanes, courts, and by-ways."[16] Even in these personal narratives the urban atmosphere leads Melville to attempt that overview necessary for his literary narrative, and literary cartography, in *Moby-Dick.*

Overview, as a technique and a general characteristic of literary narrative, is developed partly in order to deal with the problem of urban representation. Specifically, many nineteenth-century authors wanted to register the experience of the metropolis, a relatively new form of social space that was transforming social relations at the time, by capturing a "bird's-eye view" of the place. The writerly need to represent the city also correlated to the societal need to regulate it. In *Commissioned Spirits,* Arac asserts that "the chaos of urban experience fostered a wish for a clarifying overview," and the "narrative overview in fiction accompanies new techniques of inspection and centralization worked out by a growing interventionist bureaucracy."[17] While elements of this bureaucracy had been in place for a long time, the intensified urbanization of the nineteenth century called for more effective and more widespread practices of power and knowledge. In combating crime and disease, each abetted in its spread by densely packed urban populations, governmental agencies developed such techniques of surveillance that would allow for a nearly total visibility. (Michel Foucault's examination of Bentham's Panopticon in *Discipline and Punish* is the canonical point of reference for any discussion of this phenomenon.) The disciplinary society which emerges in the early nineteenth century establishes extensive, multivalenced, and continuous surveillance as its primary modus operandi, and—though it is not Foucault's desire to focus exclusively on the urban—the powers of observation are mobilized most thoroughly in the ordering of human multiplicities in the city. For the writer trying to make sense of the larger, more complex, and increasingly mobile society of the time, the technique of overview may be developed parallel to the juridical-political practices of disciplinary surveillance.

Yet the "clarifying overview" need not be understood merely as another mechanism of police-power, as anyone who has ever become lost can confirm. The clarifying overview afforded by a map, for instance, may not only bring solace (and even dissolve real panic) but also becomes an essentially epistemic device, a way of knowing a given space.[18] The distinction between itinerary and map discussed in this chapter certainly should not be taken as a distinction between freedom and coercion. To be sure, maps are always and already bound up in those power/knowledge networks which are the subject of Foucault's genealogical studies, but (as Foucault also insists) that does not mean that mapmaking is itself always and only a repressive practice. The inability to map one's position relative to a geography and social totality, as Jameson suggests, is perhaps the emblematic form of modern or postmodern alienation. And though maps may function to enforce boundaries, to monitor movements, to aid the police, to capture a given space, and so on, maps may also have liberatory uses. The map of the prison, for instance, may be of helpful to one who wishes to escape.

Walking in the City

Whereas Melville's first three books had begun at sea, *in medias mare,* as it were, *Redburn* begins on dry land. As in *Moby-Dick,* a number of early chapters are devoted to the voyager's itinerary towards the ocean and the call of the sea that draws him. There is, however, a notable difference between the first chapter of *Redburn* and the "Loomings" of Ishmael's narrative. In *Redburn,* one finds no apostrophe to the "magic" of water, to the power of the sea that magnetically draws both the urban throng and the country bumpkins. When the Ishmael of "Loomings" rhetorically asks, "Why is almost every robust healthy boy with a robust healthy soul in him, at some time or other crazy to go to sea?" (5), one is tempted to think of Redburn's first voyage. Indeed, *Redburn*'s first chapter is titled, "How Wellingborough Redburn's Taste for the Sea was Born and Bred in Him." But a quick reading of this chapter reveals that young Redburn's "taste for the sea" has little to do with the mystical power of Ishmael's water; rather, Redburn is drawn by the less mystical, though perhaps still mystifying, power of the city. To be sure, the language of Redburn (as narrator) certainly carries some romantic reveries, but these are directed, not towards the sea as such, but towards the cities to be found on the other side of the ocean:

> [M]y father, now dead, had several times crossed the Atlantic on business affairs, for he had been an importer on Broad-street. And of winter evenings in New York, by the well-remembered sea-coal fire in old Greenwich-street, he used to tell my brother and me of the monstrous waves at sea, mountain high; of the masts bending like twigs; and all about Havre, and Liverpool, and about going up into the ball of St. Paul's in London. Indeed, during my early life, most of my thought of the sea were connected with the land; but with fine old lands, full of mossy cathedrals and churches, and long, narrow, crooked streets without side-walks, and lined with strange houses.[19]

Redburn's "taste for the sea" is bred in fact by the siren call of the city. More particularly, to the "big" city: industrial and commercial ports like Liverpool and Le Havre, and the great, metropolitan capital or "global city" of London.[20]

Another, more pressing reason that young Redburn actually does go to sea—a reason both Melville and Ishmael knew all too well—is not unrelated to the urban allure: want of money. Redburn's first voyage, like Melville's own, is not a pleasure cruise but a job; he is required to find work because the family does not have the money to support him, much less send him to university. The trials related in the opening pages of *Redburn* involve his inability to pay the full cost of the ferry to New York City and his tribulations in attempting to sell an old gun (the "fowling piece") there. Like people from all over the hinterlands, Redburn goes to sea, becoming "at sea" in the city, for primarily economic reasons. The connection between money and the city is established up front, and it reflects a world in which all that is solid has melted into air, in which a noble surname carries less respect than a full purse, and in which the skeletal features of a global economic system become visible as a network, or matrix, in which the major cities functions as the nodes of power. The *bürgerliche Gesellschaft* (which could just as well be translated *urban society*, as much a civil or bourgeois society) that rescues, in Marx and Engel's words, "a considerable part of the population from the idiocy of rural life,"[21] thus exceeding the bounds of the agricultural estate, also exceeds the bounds of the national state; the "wealth of nations" will depend on the international relations among cities.[22]

Melville's America had no direct counterpart to the system developed within the European nations (or the "core" as described by Immanuel Wallerstein), in which the consolidation of state power involved the simultaneous movements outward, towards ever greater extraterritorial domains, and inward, towards greater centralization embodied by the great European capitals, the baroque metropolises, such as Madrid, Paris, London. The metropolitan center functions, at once, as the seat of an ever-more-powerful state government and as the principal agora for an expanding, international market. For a number of reasons—the system of government, the displacement of the political capital to Washington, D.C. (and away from the commercial centers of New York, Philadelphia, Baltimore or Boston), the still-dominant agricultural mode of life for the vast majority of citizens—Melville's America was, in Charles Baudelaire's words, "a country without a capital." But the urban development was well underway. In the 1840s alone, the urban population of the United States increased over 90 percent, growing three times as fast as the total population.[23] The New York City in which Melville was born in 1819 was not the same town in which he completed *Moby-Dick* years later. The burgeoning and disruptive social and economic forces that caused Melville's father, Allan Melvill, to depart from Water Street and move his family upstate had, by the mid-nineteenth century, transformed the town from "a knowable community" (in Raymond Williams's apt phrase) to a new metropolis.

In the emerging urban literature of the nineteenth century, two figures are commonly deployed in the representation of urban space: the *panorama*, a sweeping overview of the city which affords a sense of order in an increasingly chaotic social space, and of which the map is the principal representational form, and the *labyrinth*, also a form of imagined order, but in this case one that does not offer solace to one who is inside it. In its compounding of confusion and mystery within a spatial framework, the labyrinth seems to preclude any reassuring sense of overview. The labyrinth represents something unknowable; it is, if not impossible to know, then at least difficult to know, mysterious, involving a "Talismanic Secret."[24] Often, of course, the secret of the labyrinth, the purpose, is the key to escaping from it, which indicates the mood associated with the urban space. The city, like the ultra-urban figure Edgar Allan Poe discovers in "The Man of the Crowd," is a text that does not let itself be read.[25]

The breakdown of knowable communities certainly relates to that sense of alienation accompanying a general movement from country to city, a key theme in so many novels, and Marx's phrase about the "idiocy of rural life" seems to have more to do with the idiosyncrasies of rural experiences than their intelligence or intelligibility. The still largely Dutch-American New York City of Washington Irving's (and of Allan Melvill's) time was not rural; it was a commercial city, and—by the standards of the day and relative to others in the western hemisphere—a good sized one. Yet it was still what Wyn Kelley calls a "Walking City," particularly at a time when New York City (its heart, at least) was usually understood to refer to those areas between Greenwich Village and Wall Street, in which a perambulating stroller could reasonable "master his urban environment in a day's walk."[26] This knowable community, this "walking city," was becoming more and more impossible in the nineteenth century. As Raymond Williams explains,

> identity and community became more problematic, as a matter of perception and as a matter of valuation, as the scale and complexity of the characteristic social organisation increased. Up to that point, the transition from country to city—from a predominantly rural to a predominately urban society—is transforming and significant. The growth of towns and especially of cities and a metropolis; the increasing division and complexity of labour; the altered and critical relations between and within social classes: in changes like these any assumption of a knowable community—a whole community, wholly knowable—became harder and harder to sustain.[27]

The unprecedented growth of New York City during the thirty years between Melville's birth and the writing of *Moby-Dick* transformed the city, and in turn the experience of the city, almost beyond recognition, much as do the transformations of Liverpool that so disorient young Redburn, forcing him to recognize that he cannot be his father, and his father's Liverpool cannot be his. Indeed, as Redburn's experience of Liverpool demonstrates, the knowable community has all but disappeared, and the quest *to know* the city seems doomed at the outset. This inability to know the city makes up the tragedy, and much of the comedy, of *Redburn*.

When Redburn first arrives in Liverpool, he brings with him a guide-book that his father had used years before; therefore, "it had been thoroughly tested, and its fidelity proved beyond peradventure." His scrupulous study of this guide-book assured Redburn that he had acquired "an unerring knowledge of Liverpool; especially as I had familiarized myself with the map, and could turn sharp corners on it, with marvelous confidence and celerity."[28] The guide-book is titled "The Picture of Liverpool," and Redburn—partly out of naïveté, partly out of nostalgia—is utterly convinced of its entire accuracy. When the map does not bear "the slightest resemblance" to "the identical place itself," Redburn rationalizes that "this is owing to my taking a horizontal view, instead of a bird's-eye survey. So, never mind old guide-book, *you*, at least, are all right." Several pages follow in which the trusted old guide-book is proved repeatedly to be false, out-of-date, and as such no longer a guide at all. Redburn concludes to himself: "Guide-books, Wellingborough, are the least reliable books in all of literature; and nearly all of literature, in one sense, is made up of guide-books."

The narrative of Redburn's disillusionment with his much loved old guide-book introduces a lengthy chunk of the novel devoted to Redburn's experience of Liverpool. As an introduction to the urban narrative, the guide-book chapters function both as critiques of cartographic representation (insofar as a perfectly mimetic image of the space is impossible, and even "good" maps can lead one astray) and as profiles of the ever-changing world of urban experience. In a sense the two chapters devoted to the guide-book, which demonstrate its overall lack of utility and failure of representation, are themselves a kind of map, or a legend, of urban spaces.[29] The flaws of the guide-book highlight the ambiguities of the urban in Melville's literary cartography.

Coming to terms with the representational failures of the guide-book, Redburn finds himself in an urban labyrinth, forced to "follow my nose throughout Liverpool" thereafter.[30] This way of moving about the city liberates Redburn from his reliance on the guide-book (that is, from the key to the city's legibility) but, at the same time, it makes Redburn's urban itinerary all the more bewildering and alienating. The guide-book, along with its map, had presented a panoptic overview of the city, however inaccurate the resulting picture may be, that presumably enabled a somewhat detailed knowledge of Liverpool, or so Redburn had thought. Without this epistemic guide, this Ariadne's thread, Redburn finds it difficult to map the domain, and in this difficulty lie the terrors of the labyrinth.

Redburn does attempt to overcome his alienation in the face of the confusing urban milieu through a process similar to that described by Lynch in *The Image of the City*. After the guide-book and its map are proved unreliable, Redburn continues to perambulate the streets of Liverpool, attempting to produce a cognitive map his environment. This process involves, as Jameson puts it in his encapsulation of Lynch's argument, "the practical reconquest of a sense of place and the construction and reconstruction of an articulable ensemble which can be retained in memory and which the individual subject can map and remap along the moments of mobile, alternative trajectories."[31] Chapters 31 through 45 of *Redburn* are devoted to this mapping project, as Redburn tries to make sense of Liverpool on his own and without guides. Notwithstanding the time ("four weeks or more") and energy spent in this effort, Redburn does not in any satisfactory way master his domain. The city remains a mysteriously illegible text, an unknowable community. Again, like Poe's "Man of the Crowd," the city is a text of which it may be said, "*es lässt sich nicht lesen.*"[32]

In his essay "Walking in the City," Michel de Certeau establishes a different view of the urban text, in which he contrasts the image of the city (New York City, in this case) as viewed from the top of the then newly erected World Trade Center with that available to one down on the streets, amid the bustling urban throng.

> To be lifted to the summit of the World Trade Center is to be lifted out of the city's grasp. One's body is no longer clasped by the streets that turn and return it according to an anonymous law; nor is it possessed, whether as player or played, by the rumble of so many differences and by the nervousness of New York traffic. When one goes up there, he leaves behind the mass that carries off and mixes up in itself any identity of authors or spectators. An Icarus flying above these waters, he can ignore the devices of Daedalus in mobile and endless labyrinths far below. His elevation transfigures him into a voyeur. It puts him at a distance. It transforms the bewitching world by which one was "possessed" into a text that lies before one's eyes. It allows one to read it, to be a solar Eye, looking down like a god. The exaltation of a scopic and gnostic drive: the fiction of knowledge is related to this lust to be a viewpoint and nothing more.[33]

De Certeau acknowledges the desire to see the city as a whole, noting that medieval artists created utopian images of the city that no human eye could have seen, but he associates this "scopic drive" with a baleful totalization. But de Certeau does not take into account the apparent need, from the standpoint of the individual subject, to map his or her *Lebenswelt* in this way. On the contrary, de Certeau believes that this sort of overview is a ruse, or a conceit, and that the true "practitioners of the city"—as opposed to the "voyeurs" high above, literally or otherwise—are down on the streets: walkers, *Wandersmänner* who resist the panoptic totalization of the map. Somewhat in the fashion of Baudelaire's panegyric to the *flâneur*, the intellectual poet strolling idly in the city who can become "a *kaleidoscope* equipped with consciousness,"[34] de Certeau sees in the "window shopper" a more liberated and liberatory figure of urban modernity.

In de Certeau's argument, the division between description and narration, between image and text, reappears. The observer from on high acquires or, rather, attempts to acquire a synoptic and panoptic image of the multiplicities and mobile realities below. In this way, the "bird's-eye" viewer attempts to "freeze" the city into a stable and homogeneous image. Conversely, a city-walker is the true writer or narrator of a text which is constantly unfolding and in which he is himself inscribed. The city-walker's activity, his or her spatial practices, according to de Certeau, disrupt the stable image or description, and instead produce a narrative, a "long poem."

> The long poem of walking manipulates spatial organizations, no matter how panoptic they may be: it is neither foreign to them (it can only take place within them) nor in conformity with them (it does not receive its identity from them). It creates shadows or ambiguities within them.[35]

De Certeau's reading of this "long poem of walking" and its effects on the panoptic organization of space recalls Georg Lukács's critical contrast of realism and naturalism in "Narrate or Describe." Lukács notes that the distinction between realist narration and naturalist description is identical to the distinction between experiencing and observing, and that these are two "basically divergent approaches to reality." Realism is motivated by a participating narrator, one who both creates the story and is part of it. As such, the narrator of the realist text is a free agent, one who can actively determine the events unfolding in the narrative, unlike the naturalist narrator, who can merely describe situations, characters, and events observed at some distance. "Description contemporizes everything. Narration recounts the past. One describes what one sees, and the spatial "present" confers a temporal "present" on men and objects."[36] Lukács views the latter with disdain, of course, but the cartographic activity involved in describing is no less present in narrating, and often the two overlap. The recounting of events and places involves a dynamic mapping practice that is irreducible to either the static map or the transient itinerary.[37]

Redburn's initial reliance on his map and guide-book, and his subsequent attempt to map the space of Liverpool without their assistance, conjures up a different image from de Certeau's "long poem of walking" or Lukács's critique of description as opposed to narration. Redburn's angst and disorientation derives largely from his inability to achieve that narrative overview that will allow him to "read" the city. Clearly Redburn is not in the position of de Certeau's free agent narrator, even though *Redburn* is a personal narrative whose apparent authority rests almost entirely upon the narrator. While he is the principal character and central figure in the narrative, Redburn seems to have more in common with Lukács's naturalist observer, at least insofar as his agency is concerned. Unlike Lukács's observer, however, Redburn is unable to achieve that scientific distance from the reality he describes, since he is very much in the midst of what is going on. Redburn is constantly "acted upon," and when he tries to act for himself, the results are usually tragically humorous. He is not the agent who can determine the events of which he is a part; rather, Redburn is swept up in a whirlwind of forces over which he has little or no control. When Redburn does try to act, to affect his environment (as when he attempts to get help for the homeless woman and child in Launcelott's Hey), he is met with disappointment at best, and ridicule more often than not. Even where the scene is less horrible, Redburn is much more subject to the environment than he is a subject acting within it, as is dramatized by his exhilarating entry into the metropolis of London:

> As we rattled over boisterous pavements, past splendid squares, churches, and shops, our cabman turning corners like a skater on ice, and all the roar of London in my ears, and no end to the walls of brick and mortar; I thought New York a hamlet, and Liverpool a coal-hole, and myself somebody else: so unreal seemed everything about me. My head was spinning round like a top, and my eyes ached with so much gazing: particularly about the corners, owing to my darting them so rapidly, first to this side, and then that, so as not to miss anything; though, in truth, I missed much.[38]

The breathlessness of the narrative voice here parallels the hyper-hurly-burly of the metropolitan scene. If Redburn were unable to form an adequate cognitive map his domain in Liverpool, a mere "coal-hole," then he has no chance whatsoever in London, the city that he had most looked forward to seeing and that he would have the least opportunity to explore.

The urban itineraries in *Redburn* reveal an element lacking in de Certeau's account of the city-walker: the alienation of the individual subject who cannot adequately map his or her environs. If the view from the top of a skyscraper creates an illusory sense of mastery, it does at least render the text of the city readable, if only through the provisional creation of a spatial and allegorical code that cannot actually record "the identical place itself." And, contrary to de Certeau's implications, this legibility opens the urban text up to multiple interpretations utterly unavailable to the non-readers (though they may be "authors") on the street. That is, the image of the city as projected through some sense of overview is not necessarily the only image available; rather than ignoring the "devices of Daedalus," the "scopic and gnostic" viewer attempts to make sense of those devices. Furthermore, de Certeau's argument does not adequately acknowledge the degree to which the city-walkers are also, like Lynch's city-dwellers, attempting to map their urban environment in ways that may lead them to a kind of speculative overview (even when they do not ascend the position of a solar Eye).[39] We who are in the labyrinth would undoubtedly like to see a map of it. Failing that, we will attempt to map it ourselves, using—like the *bricoleur* of Claude Lévi-Strauss—whatever devices are at hand.

Melville's comic depiction of Redburn's urban itinerary offers a more light-hearted view of the labyrinth, underscoring the bewilderment of being lost in a big city without emphasizing the pathos. After *Moby-Dick,* with its newly discovered techniques to enable a clarifying overview and its comprehensive mappings, Melville returns to his urban cartography with *Pierre, or, the Ambiguities* and "Bartleby, the Scrivener," providing as much pathos as could be desired. With these works, Melville approaches the limits of his literary cartography, as the wide expanses of the world system are reconfigured in a descent into the sepulchral and claustrophobic heart of the metropolis.

Chapter 8

The Ambiguities of Place: Local Narrative and the Global City

The city that proved so bewildering to the young Wellingborough Redburn is not merely a labyrinthine space but also a unique place. The urban locale, like the distinctive regional spaces depicted in nineteenth-century American local narrative, presents an eccentric experience, one out of the ordinary or familiar, often at odds with the mainstream image presented in national narrative and less likely to be incorporated into national space than those exotic zones explored in personal narratives. Yet, unlike those extraterritorial places overseas or those tucked away regions like Dutch New York or the old Southwest, the city is literally central, lying at the heart of the nation-state and forming the key node of the multinational world system. The city, or more particularly, the metropolis, is thus a special locale; it provides the grounds for a local narrative, but one whose effects are different from the others that I discussed in Chapter 2. For whereas most local narratives presented out-of-the-way places that were unique and inassimilable to national narrative, primarily because they so clearly exhibited their uniqueness and eccentricity, the urban *topos* is ambiguous and central. It is not out-of-the-way, but it is not clear either. The metropolitan cityscape will undermine the image repertoire of American national narrative and also the cartographic project of making clear the diverse spaces of the world. The ambiguities of the urban demonstrate the difficulties, and perhaps ultimate failure, in attempting the map the world system.

In this chapter I will examine the problem of the global city in Melville's mapping project by looking at two post-*Moby-Dick* works: *Pierre* and "Bartleby, the Scrivener." I do not provide readings of these narratives, exactly, but rather invoke them as exemplary interventions in Melville's struggle to incorporate the global city into the map of the postnational world system. Each tale functions in very different ways, and each is rich and complex enough to be worthy of an entire study. Yet here I am primarily interested in a few key elements of each. First, in *Pierre,* Melville's first venture after the triumphant-though-catastrophic global literary cartography that was *Moby-Dick,* Melville deliberately establishes his Dantean, Faustian, and Hamlet-like tragedy as an allegory of contrasting spaces. The country versus the city, which does not preclude the antagonism of the nation versus the city, becomes the thematic underpinning of both Pierre Glendinning's downfall but also the collapse of a certain national image. Second, in "Bartleby," the admixture of commerce, professionalism, and reading form elements of a local narrative that is necessarily urban while also taking ambiguity to the level of inscrutability. That is, with "Bartleby," the local narrative subject is not merely eccentric, but unreadable. In so doing, Melville reaches the limits of his own mapping project.

The global city, a metropolis that is no longer (like the great European capitals) the center of a national cultural space, but is a key nexus in a transnational or postnational world economic system. My use of the term "global city" in the context of Melville's work is, of course, anachronistic. Saskia Sassen coined the term and limited its initial use to three global cities (New York, London, and Tokyo), in order to register the distinct type of metropolis in the commercial and financial world under globalization.[1] It would be a stretch, as well as inelegant, for me to use the term "proto-global city" to describe Melville's New York, but even the mid-nineteenth-century New York had become a center of finance (over and above manufacturing, for instance), and the unique situation of New York as a "world city" rather than a national capital imbues the city with an air of globalization that precedes the actual economic globalization of the late twentieth century. Though merely a small island in the sea of commerce, New York is a symbol of the swirling world system of international finance. "There now is your insular city of the Manhattoes, belted round by wharves as Indian isles by coral reefs—commerce surrounds it with her surf."[2] In New York, the emergent global city figures forth the representational crises that trouble Melville's literary cartography. The global city becomes a space of disruption, overwhelming of the national and challenging clear legibility and stable meanings; in this space the postnational forces ultimately emerge.

The metropolis or "global city" can be visualized as an alternative space to the imagined community of the nation-state. Indeed, the country-versus-city divide can be seen as an antagonism, not merely between the rural and the urban, but between the state or nation and city; this is particularly the case in the popular imagery of the American "values," national values which are also more commonly associated with rustic life. (Every election season seems to provide an occasion to reiterate this view of the nation, as various self-proclaimed representatives of "real America" champion small towns or rural locales, implicitly or explicitly stating that the urban locale is somehow un-American, notwithstanding that fact that the overwhelming majority of the population—some 80%, based on the 2000 Census—dwells in metropolitan areas.) But, of course, the coordinates of the country-versus-city model of thinking about society have never required specific numbers or straightforward demography, and the city/country divide has increasingly moved beyond the spatial scope of the national. As Raymond Williams puts it, "a model of city and country, in economic and political relationships, has gone beyond the boundaries of the nation-state, and is seen but also challenged as a model of the world."[3]

A Docile Country

A few months after the publication of *Moby-Dick,* Melville, hard at work on his next novel, wrote to Sophia Peabody Hawthorne, expressing his gratitude for her "highly flattering letter" in which she praised the whaling epic. Melville expresses amazement, both in her satisfaction with the book and in her keen perspicacity which allows her to discern allegories that he had not intended, and he notes that "some men have said that they were pleased with it, but you are the only woman—for, as a general thing, women have small taste for the sea." Melville then makes this promise: "But, My Dear Lady, I shall not again send you a bowl of salt water. The next chalice I shall commend, will be a rural bowl of milk."[4]

The dichotomy between the sea and the land is here gendered; the former belonging to men and the latter to women, but Melville's metaphor goes further.[5] By associating the softer, more comfortable region with the rural, Melville is distinguishing not between sea and land, but between other things: a non-rural precinct unspecified here and the rural, from which one tastes the salubrious and above all *homely* milk. Given what we know will happen in *Pierre, or, The Ambiguities,* such a comment can only be viewed as ironic. *Pierre*'s "rural bowl of milk" quickly sours into a scarcely potable if not downright poisonous liquor by the end of the story. And, significantly, the souring process correlates directly to the narrative's geographic movements as the story arcs from a pastoral idyll to an urban nightmare. After *Moby-Dick,* the city itself becomes a figure for the inscrutable system embodied in the whale.

As rapid urbanization was transforming daily lived experience in the mid-nineteenth-century United States, it was also transforming the "America" celebrated in the dominant national narrative. The image repertoire of Americanist State philosophy depended heavily on a rustic or agrarian premise: a Virgin Land, an errand into the wilderness, and an individual "Adamic" subject. The massive concentration of populations in the large (and steadily growing) cities challenged the pastoral ideal and the national image associated with it. As such, much of the urban literature of the nineteenth century presents the city as an almost foreign or exotic locale, and more often than not, the city represents a kind of dystopian reality. The personal narrative in *Redburn* only hints at this. *Redburn* mobilizes many of the elements of popular urban literature of its time, but it stops short, again and again, of replicating the results. As a tale of lost innocence and gained experience, which on the face of it *Redburn* is, the novel curiously fails to show that the Redburn at the end of the book is significantly different from the Redburn who first left home to seek his fortune. He has obviously not made his fortune (as in novels of social ascension), but neither has he fallen into ruin (as do the heroes and heroines in the immensely popular romances of the day). *Redburn* is not an odyssey which lends itself to morally didactic purposes; by the end it appears to have been little more than the scattered remembrances of a largely unremarkable voyage. But by its focus on the urban as a place of bewildering inscrutability, *Redburn* does present a counter-image to the America of national narrative.

In Chapter 2, I discussed how local narrative, consciously or otherwise, served to undermine the aims of national narrative by insisting on the uniqueness of particular locales. Local narratives tended to be detailed sketches of eccentric people and places; the eccentricity was a key aspect of their appeal. The reader, who might be representative of a mainstream (and perhaps emerging national) culture, could marvel over, laugh at, or simply look upon the strange characters and odd sites of subcultures outside of the main currents of public life. The detail of these narratives made the off-centered topics knowable, as the readers could point to their subjects and see clearly just how outside-of-the-mainstream they actually were. For example, the backward country folk of Southwest Humor—such as the youth in Longstreet's "The Fight"—can be presented in such detail that the reader may condescendingly laugh at the rustic ignoramus while also understanding the foreign culture (here, backwoods Georgia) to be an eccentric locale inside "America." Like personal narratives, local narratives functioned as a way of making known the relatively unknown or exotic experiences, but unlike personal narratives, local narratives tended to keep these at a distance rather than incorporating them into a national cultural image. That is, the spaces of local narratives, though known, remain eccentric.

The city, as a locale for local narrative, operates in a somewhat different register, both because the urban space need not be limited to a particular region and because the eccentric experience of urban life was not literally off-center, but often lay in the very heart of the public sphere. Indeed by its very nonrurality, the city becomes, paradoxically, a central-but-eccentric space within the American national geography. The strangeness of the urban *topos* is heightened by its dominance, and the image of the city remains one that causes fear and trembling while also operating as the model for social organization itself. It also helped to establish an attitude toward oneself and one's community that made possible literary narrative's intensive psychological force: "The pressure of urban crowds and the split between home and work both helped to establish a heightened sense of personal interiority and a need to define and cherish particularity."[6] The new conditions of work, the increasingly specialized divisions of labor, and the emergence of professions were associated with the rise of the city, and contributed to the undermining of the ideals of national narrative. The country-versus-city divide—that is, the pastoral or rustic versus the urban—thus becomes also an antagonism between nation (or "country") and city. In the central-but-eccentric urban space, the local, personal, national, and literary narrative forms come to a head in a chaotic conflict over how the self, the community, the nation, and the world could be imagined.

Melville, without directly endeavoring to produce either local narratives (though "Bartleby," described by reviewers as "Poe-esque," might qualify) or novels of urban exposé (as, for example, in Dickens), addresses this country/city divide at its most formulaic in *Pierre, or, the Ambiguities*, in which both pastoral sentimentalism and urban sensationalism are mobilized. His "rural bowl of milk," as he promised Sophia Hawthorne, dredges up the most gothic elements of popular urban potboilers, right down to its final scene in "the Tombs"—which is, of course, also where Bartleby, the scrivener, ends his days. The popular formula of the well-bred country gentleman's fall from grace in the metaphysically troubled space of the metropolis appears in very simple terms. The country is a space of tradition, of stable meanings, of beneficent patriarchy (here, with a matriarch), of a temporal languor or even immobility, of comfort. This is the domestic space of the home, the homely or familiar. The city, on the contrary, is reckoned as the space of constant upheaval, where things are not always what they appear and where things change, evanesce, reappear transformed; it is a place of speed, of the jostling hurly-burly of commerce and crowds, of ambiguity, with its unstable semic field and multiples meanings. In its speed, movement, and power, the growing city of the nineteenth century is coupled—as actually cities are literally linked—to another great nineteenth-century development, the railroad, where sedate and sedentary experience dissolves into a dizzying, perhaps terrifying, array of images torn from their "natural" coordinates.[7] The antipodal image of the domestic and docile country, the city does not feel like "home." Urban angst has to do precisely with the unhomely nature of the place. As Heidegger points out, "In anxiety one feels "*uncanny*" [*unheimlich*] [. . .] But here "uncanniness" also means "*not-being-at-home*" [*das Nicht-zu-hause-sein*]."[8] In *Pierre*, this uncanny anxiety that is also a form of homelessness is dramatized in Pierre's movement from the country to the city.

Pierre opens with a typical description of the country/city divide, one which, not surprisingly, comes down firmly in favor of the country. The very first line suggests the difference in perspective between a city-dweller and a rustic: "There are some strange summer mornings in the country, when he who is but a sojourner from the city shall walk forth into the fields, and be wonder-smitten with the trance-like aspect of the green and golden world."[9] A few pages later, the reader learns that it "had been [Pierre's] choice fate to have been born and nurtured in the country, surrounded by scenery whose uncommon loveliness was the perfect mould of a delicate and poetic mind"; so significant is this point that the narrator repeats it later in the chapter, this time underscored.[10] While Pierre has the added advantage of being conversant with urban culture, having visited the city annually since youth, Book I of the novel asserts again and again that the country-life has a distinct advantage over life in the city, as in the following observation:

> For it is to be observed, that while in other countries, the finest families boast of the country as their home; the more prominent among us, proudly cite the city as their seat. Too often the American that himself makes his fortune, builds him a great metropolitan house, in the most metropolitan street of the most metropolitan town. Whereas the European of the same sort would thereupon migrate into the country. That herein the European hath the better of it, no poet, no philosopher, and no aristocrat will deny. For the country is not only the most poetical and philosophical, but it is the most aristocratic part of this earth, for it is the most venerable, and numerous bards have ennobled it by many fine titles. Whereas the town is the more plebeian portion.[11]

The panegyric to the country, and the concomitant disparagement of the city, at the beginning of *Pierre*, sets the stage for a formulaic tale of urban dissolution and destruction to follow, but it also enables Melville to further his critique of American State philosophy. The rural landscape, in this case Pierre's Saddle Meadows, comes to be associated with the America of national narrative, and the city (New York City) in turn becomes a space of the new, postnational space that stands in stark contrast to the exemplary, and exceptionalist, rustic American ideal.

Of course, the very terms *country* and *city* offer somewhat ambiguous labels for the various kinds of social experience associated with them. In the passage above, for instance, Melville conflates the two most common uses of "country" by invoking nation-states ("in other countries" or "Europeans") alongside rural domains ("the country [rather than the city] as their home"). By contrasting the pastoral and the metropolitan in terms of where the wealthy or the "finest families" live and of what the poets and philosophers say, Melville also suggests a constellation of space, power (political and economic), and history. Pierre's Saddle Meadows affords an exemplary model of both the aristocratic European estate and the ideological Americanist mini-republic. For the Glendinning family seems emblematic of "Americanness." Notwithstanding the rhetoric of revolutionary American nationalism, which tended to emphasize the forward-looking movement and general paucity of moldering old history—e.g. with the image repertoire of the Virgin Land, an American Adam and an errand into the wilderness that fostered the nation of ever-expanding futurity, which does not, as does Europe, have the great burden of the past—national narrative was deeply bound up in establishing a national history, albeit a mythic one. Saddle Meadows, for *Pierre*, is densely saturated with such history. As in Hawthorne's house of the seven gables, in which the history of dead generations weighs like a nightmare on the brains of the living

Pyncheons, Pierre's ancestral home is burdened with a past, but here it is a history that seems coterminous with that of the nation. Pierre's social condition is introduced by the assertion of "the great genealogical and real-estate dignity of some American families," and his family's aristocracy is established less by wealth and longevity—"the English farmer Hampdens were somewhat the seniors of even the oldest Glendinning"—than by association with the birth of the republic: Pierre's grandfather, a Major-General in the Revolutionary War, provides the credentials. Saddle Meadows thus represents, in a sense, the law of the fathers, and more particularly, of the Founding Fathers, the patrilineage not only of Pierre Glendinning, but of the American nation itself.[12]

Saddle Meadows may thus stand in for the "country" in both senses, as the rural domain and the national. Saddle Meadows, as in the more general notions of country-life, is a place of stability and docility. This last word is significant, and Melville's emphasizes this significance for the reader. At the conclusion of the first chapter, when Mary Glendinning, Pierre's mother and the heir to the family's national-aristocratic mien, regards her son, she uses the word "docile" no fewer than nine times in a single paragraph.

> "A noble boy, and docile"—she murmured—"he has all the frolicsomeness of youth, with little of its giddiness. And he does not grow vain-glorious in sophomorean wisdom. I thank heaven I sent him not to college. A noble boy, and docile. A fine, proud, loving, docile, and vigorous boy. His little wife [Lucy], that is to be, will not estrange him from me; for she too is docile,—beautiful, and reverential, and most docile. Seldom have I known such blue eyes as hers, that were not docile, and would not follow a bold black one, as two meek blue-ribboned ewes, follow their martial leader. How glad I am that Pierre loves her so, and not some dark-eyed haughtiness, with whom I could never live in peace; but who would be setting her young married state before my elderly widowed one, and claiming all the homage of my dear boy—the fine, proud, loving, docile, vigorous boy!—the lofty-minded, well-born, noble boy; and with such sweet docilities! See his hair! His does in truth illustrate that fine saying of his father's, that as the noblest colt, in three points—abundant hair, swelling chest, and sweet docility—should resemble a fine woman, so should a noble youth."[13]

The "docile" monologue follows a seemingly light-hearted, but ultimately ominous, conversation between mother and son concerning Romeo's sad fate. "It was his own fault," says Mary Glendinning, "He was disobedient to his parents."[14] As William Spanos points out, however, the "world" demands something else.

> It is not obedience as such, which connotes the overt presence of an economy of constraining rules and regulations—and thus, in its visibility, is vulnerable to the possibility of insurrection. It is docility, a comportment towards the

> world, enabled by what Louis Althusser has called "the interpellation of the subject" and the related mechanisms of normalization, that assume its constraints to be natural, the way things actually are, and thus assure peace and confidence: "so be it!" The word "docile"—in tension with a peace-threatening barbaric energy—reverberates throughout this self-consciously "pleasant" conversation like the ominous clanging of a bell, a process that culminates in the concluding anxious thoughts of Pierre's mother. These, not incidentally, are articulated beneath the supervising gaze of the memorial portrait of her father, "grand old Pierre," a "hero," we have been previously told, of the American Revolution, and a slaveholder, albeit a benign one.[15]

Docility, as in Foucault's inquiry into "docile bodies," is the necessary condition for well-regulated government, and a prerequisite for (as well as a continuing effect of) modern state power. In Althusserian terms, this is "ideology," but it is perhaps more clear in its nuanced saturation of every corner of lived experience in Antonio Gramsci's model of "hegemony," by which the relations of power become almost entirely invisible to those caught up in them.[16] It is surely no accident that, in Mary Glendinning's "docile"-laden paragraph, the final lines link the docility of "the noblest colt" with that of "a fine woman" and "a noble youth." Add to this the reference to slaveholding, and that the slaves "all of them, loved grand old Pierre, as his shepherds loved old Abraham,"[17] and the repeated sounding of the word "docile" acquires and retains a more sinister character.

Furthermore, the military aspect of the Glendinning family's nobility is not unrelated to the importance placed on docility by a matriarch who had a martial air (after her monologue on docility, "her glad proud eye" resting on "the old General's baton," and "now she looked the daughter of a General, as she was").[18] As Foucault has made clear, in terms that resonate with Mary Glendinning's dream of a perfect family, in "the military dream of society" (as contrasted to the philosophers' eighteenth-century, utopian view), the

> fundamental reference was not to the state of nature, but to the meticulously subordinated cogs of a machine, not to the primal social contract, but to permanent coercions, not to fundamental rights, but to indefinitely progressive forms of training, not to a general will but to automatic docility.[19]

The docility of the young Pierre, of Lucy Tartan, of the servants, indeed, of the entire culture of Saddle Meadows, conforms to this "military dream of society," a well-regulated state in which rules need less to be enforced as to be accepted as the natural state of affairs, or, again, what Gramsci referred to as *hegemony*.

The country, in *Pierre*, is thus conceptually linked to the nation-state (or the country). The crisis at the heart of the novel, then, is less a personal "crisis of conscience" or an individual fall from grace than a rupture within American national culture itself. Spanos has noted that Melville's sustained effort in *Pierre* "to think his estrangement from the American homeland in the context of the history of Old World philosophy culminates in the explosion of the myth of American exceptionalism."[20] The postnational force of *Pierre* is partly displayed in its movement from the country, in multiples senses, to the city. The crisis that leads to Pierre's expulsion from the Edenic Saddle Meadows discloses an emerging, postnational culture, whose incipient forms are located within the metropolis.

Unlike *Redburn*, *Pierre* offers little description of the city itself. Rather, the city has an atmospheric reality, which pervades the second half of the novel. The gloomy, Poe-esque environment reveals itself the moment that Pierre's stage-coach enters the city limit. The language describing Pierre's arrival in the city reinforces and reiterates that of the "Light and Gloom" chapter, the chapter in which Pierre, drawing from his studies of Dante and Shakespeare, is suffocated "in flame and smoke" by the "fiery floods in the Inferno, and all the rolling gloom in Hamlet,"[21] and in which he makes his fatal resolution. In the dark metropolis, even what light there is only accentuates the pervasive gloom:

> when the coach gained the top of the very gradual slope running toward the obscure heart of the town, and the twinkling perspective of two long and parallel rows of lamps was revealed—lamps which seemed not so much intended to dispel the general gloom, as to show some dim path leading through it, into some gloom still deeper beyond—when the coach gained this critical point, the whole vast triangular town, for a moment, seemed dimly and despondently to capitulate to the eye.

The overall darkness of the city as it first comes into view anticipates the key locations of the novel's urban half, the dark Church of the Apostles and the dank, subterranean Tombs. This darkness is itself a symbol of the illegibility of the urban experience. Kept "in the dark," the reader—Pierre himself—cannot read the city, and the mapping project that proved so difficult for young Redburn is all the more impossible for Pierre.

In a number of ways, *Pierre* is not really an urban novel at all; the city plays no direct role in the events that unfold, and, but for the manipulation (by Melville) of gothic conventions and urban sensationalism, Pierre's downfall may just as well have occurred elsewhere, in a small town as easily as in a burgeoning metropolis. Pierre, notwithstanding his flight from Saddle Meadows, carries the "country"—with all of its ideological freight—within himself. By his tragedy, Melville invites us to consider the ends of Americanist ideology, with its pastoral-national imagery: the American Adam founders on the shoals of the metropolis; the Virgin Land is revealed to be teeming with life, dense with history; and the Errand into the Wilderness becomes a bewildering journey through the labyrinth.

The city, in *Pierre*, becomes the place of this demystification, even as it is not directly stated. This is not to say that, as in *Redburn*, the city is not also itself mystifying. But the "America" presented by national narrative and figured forth the docile tranquility of the rural scene has little chance of survival (at least, not without a great deal of alteration) in this ambiguous, urban environment. The metropolis, which at first seemed to be merely a part of the nation-state and which had been a focal point of the state's centripetal force, now appears to have gone beyond the state altogether. In Melville's presentation, the protocols of reading the spaces of the nation-state are inadequate to the task of reading the city. And, where the city becomes illegible, so too does the overall cartographic project founder. In "Bartleby, the Scrivener," an original text that does not allow itself to be read becomes emblematic of that cartographic anxiety that makes one feel that the world cannot be mapped. The center does not hold.

An Illegible Text

Along with a few other sketches and tales published in between 1853 and 1856, "Bartleby, the Scrivener" is about as close to the local narrative genre as Melville produces. It is, famously, his first short story (not counting the "Town-Ho's Story" in *Moby-Dick*, which had been also published separately as a means of advertising the novel), and Melville pointedly subtitles it "A Story of Wall Street." That is, the tale is intended to describe a place and events proper to that place. *Wall Street*, in 1852 as now, is itself a synecdochic figure for a global system of high finance, for international markets and flows of currencies for stocks and bonds—and in the twenty-first century, perhaps infamously, for credit default swaps and exotic financial derivatives. On Wall Street the familiar national imagery breaks down entirely, and the map of the spaces of America becomes more and more illegible or uncertain. If the bustling city seemed to bewilder Redburn, and the gloomy urban atmosphere plunged Pierre into an infernal death spiral, the global city of advanced and advancing capitalism fosters a different sort of confusion and terror: the fear of the utterly inscrutable.

"Bartleby" conforms nicely to many of the conventions of local narrative, with a detailed sketch of a quirky, odd, and eccentric culture (that of law clerks), a specific study of a particularly odd case (Bartleby himself), and a narrator speaking as a kind of translator to a readership that might not otherwise understand the subculture presented. The twist that Melville puts on the genre, one that will render the tale a literary rather than local narrative, is that the scene and characters depicted cannot be known, regardless of the detail of presentation and elaboration. Indeed, unknowability is the main theme of "Bartleby," and the story is less about learning who Bartleby is than about the limits of reading entirely. That this inscrutability is set forth as "A Story of Wall Street" is meaningful; it brings the mapping project Melville undertakes in his early works full circle, as the downtown region of his own birth becomes the symbol of an unmappable world system.

Notwithstanding the title, "Bartleby, the Scrivener" is really about the narrator rather than his inscrutable, short-term employee. Bartleby is the subject of the narrator's case study, but—as the reader discovers—little is actually learned of the man. Bartleby is a mystery when he first arrives on the scene, becomes more mysterious still, and remains a mystery after his death in the Tombs. He is, like Poe's man of the crowd, a text that does not permit itself to be read.[22] The tale is an exercise in reading, in showing how the narrator attempts to make sense of this illegible text. That he will fail in his attempt is not accidental, and Melville's most famous short story highlights the limits of interpretation and of representation. In the context of his literary cartography, "Bartleby" is a sign of defeat, a retreat from Melville's attempt to map the world system, and a recognition of the power of the unknowable.

The narrator describes himself as "one of those unambitious lawyers who never addresses a jury, or in any way draws down public applause; but in the cool tranquility of a snug retreat do a snug business among rich men's bonds and mortgages and title-deeds." In general, many readers find him fatuous and superficial, especially as he delights in his acquaintance with John Jacob Astor (undoubtedly the richest man in the United States at the time), a man whose name he loves to pronounce repeatedly, "for it hath a round and orbicular quality to it, and rings like unto bullion."[23] But, contrary to the image presented in the crude Marxist reading of Bartleby as a victim of the capitalist lawyer, there can be little doubt of the lawyer's overall affability, good nature, and genuine concern for his employees. The lawyer's office is populated with three colorful stereotypes, known only by their Dickensian nicknames, Turkey, Nippers, and Ginger Nut, who are quirkily eccentric, though inoffensively so, and thus representative of the beings that inhabit so many local narratives. The lawyer's recent appointment as Master in Chancery increased his "original business—that of a conveyancer and title hunter, and drawer-up of recondite documents of all sorts," requiring him to hire another clerk. At first, far from standing out, Bartleby seems almost too normal for this workplace, a model scrivener, doing "an extraordinary quantity of writing," albeit doing so "silently, palely, mechanically." A change occurs when, having been asked to "read the copies" (i.e. compare the copied document to the original to assure that they are identical), Bartleby responds with the famous phrase for which the character and the story is known: "I would prefer not to."[24]

Over the course of the tale, this phrase will recur some twenty-two times. In many readings of the story, it stands as Bartleby's (and Melville's) "great refusal," a rebellious denunciation of the working and living conditions of mid-nineteenth-century American society; and it constitutes what some critics view as the great refusal, a refusal to work and to live in the society. In Leo Marx's famous interpretation, "Bartleby" is an allegory of the role of the literary writer in a society that seems to have no place for his literary art.[25] That is, Melville refused to, or *preferred not to,* produce "copies," imitations of popular novels and stories, as opposed to "original" writings like *Moby-Dick* or *Pierre.* The relative failure of *Moby-Dick* and the absolute failure of *Pierre* are what lead Melville to magazine writing as a way to earn a living, and "Bartleby, the Scrivener" is the first story he publishes. For Marx, Bartleby represents Melville's hopelessness in the aftermath of his failed literary narratives, and the silence of Bartleby reflects the estrangement of Melville, and of the literary artist in general, from his fellow man. Some critics read "Bartleby" as a direct critique of the sociopolitical conditions of capitalist society. James Wilson, for instance, notes that Bartleby's "alienation results from the dehumanizing experience of Wall Street, from the prison of his socioeconomic system," and Louise Barnett sees in Bartleby a "victim of and protest against the numbing world of capitalistic profit and alienated labor."[26] This seems to me to be a real stretch, as it establishes Bartleby as an everyman rather than an exception; if this is what dull, bureaucratic, machine-like labor does to Bartleby, then why is he unique? No. In his paean to "the meanest mariners, renegades, and castaways" in *Moby-Dick,* Melville presented his bona fides in establishing himself as, in the words of C. L. R. James, "the representative writer of industrial civilization."[27] Bartleby is something else, not merely another worker; he is a text that does not let itself be read, and to the extent that he is revolutionary, it is because one cannot make sense of him.

True, Bartleby, who had previously produced extraordinary amounts of copying, "prefers not to" read those copies. Eventually, he "prefers not to" write anymore, and by the end he "prefers not to" do anything anymore. Yet Bartleby's behavior does not really constitute a refusal to work or rebellion against the capitalist system. As Gilles Deleuze has pointed out, Bartleby does not, in fact, say "No" to the lawyer, and the careful language used is quite distinct from an outright refusal.

> The formula, I PREFER NOT TO, excludes all alternatives, and devours what it claims to conserve no less than it distances itself from everything else. It implies that Bartleby stop copying, that is, that he stop reproducing words; it hollows out a zone of indetermination that renders words indistinguishable, that creates a vacuum within language. But it also stymies the speech acts that a boss uses to command, that a kind friend uses to ask questions or a man of faith to make promises. If Bartleby had refused, he could still be seen as a rebel or insurrectionary, and as such would still have a social role. But the formula stymies all speech acts, and at the same time, it makes Bartleby a pure outsider to whom no social position can be attributed. This is what the attorney glimpses with dread: all his hopes of bringing Bartleby back

> to reason are dashed because they rest on a *logic of presuppositions* according to which an employer "expects" to be obeyed, or a kind friend listened to, whereas Bartleby has invented a new logic, a *logic of preference*, which is enough to undermine the presuppositions of language as a whole.[28]

Bartleby is not a rebel, fighting a crass, commercial, unpoetic society. Bartleby is even more of an outsider than the alienated artist; he is completely outside the system of language itself. He does not long to produce original documents instead of copies (like Nippers, who oversteps his professional bounds by trying to practice law). Rather, he *prefers* not to do anything at all. As I have argued elsewhere, "Alienated from society, from his fellow man in the office, from social conventions of behavior, and from language itself, Bartleby literally does not make sense."[29]

Once again, Bartleby is a text that does not let itself be read. As the narrator makes clear, Bartleby is not a typical scrivener, and neither is he merely the odd or quirky type one finds in local narratives. By having the Bartleby's story told by the lawyer, who concedes that he knows very little about him, Melville accentuates the originality and the inscrutability of the character. Whereas an omniscient, third-person narrator might be able to reveal more about Bartleby, and any "original source" might add useful information, the lawyer is attempting to make sense of the situation along with the reader. In "Benito Cereno," by contrast, Amasa Delano is not the narrator, so his confusion or misreading of the texts before him—Babo's elaborate charade and bewildering scenes aboard the *San Dominick*—can be judged by a reader who has a slightly different vantage. In "Bartleby," the reader may surmise that the narrator deludes himself, but the reader can gain no further knowledge.

Throughout the tale, the lawyer is perplexed. He casts about, searching not only his own considerable personal experience with law clerks but also literature and philosophy, in the hope of finding a clue to Bartleby's queer "preference." The lawyer realizes that Bartleby does not *actively* refuse to "read the copies," but (as Deleuze suggests) he cannot understand the logic of Bartleby's formula. For example, when he asks Bartleby to go to the post office and Bartleby responds with "I would prefer not to," the lawyer presses him: "You *will* not?" Bartleby responds, "I *prefer* not." Melville's own italics emphasize the distinction.[30] The lawyer is so puzzled that he consults "Edwards on the Will" and "Priestly on Necessity," but philosophy does not help him to understand Bartleby better; rather it puts his own condition in terms he can understand. "Gradually I slid into the persuasion that these troubles of mine, touching the scrivener, had been predestinated from eternity, and Bartleby was billeted upon me for some mysterious purpose of an all-wise Providence, which it was not for a mere mortal like me to fathom."[31] That is, instead of making Bartleby more knowable, the lawyer tries to makes himself more content with his own ignorance.

In revealing a rumor about Bartleby's prior experience in the Dead Letter Office, the lawyer makes a final attempt to read the inscrutable text. But the job of sorting "dead letters" does not really explain Bartleby; it merely adds another datum to the mysterious figure: presumably, Bartleby had handled letters that could no longer be read by their intended recipients; as a scrivener, he also handled "dead" letters, copies of copies. At the Tombs, the "grub-man" mistakes Bartleby for "a gentleman forger," which, of course, he is—a scrivener, like a forger, makes copies of originals. The lawyer says that he "was never socially acquainted with any forgers,"[32] that he does not know such people. Indeed, Bartleby is profoundly *unknowable* and *unreadable*.

As a story of Wall Street, "Bartleby" registers a powerful sense of novelty, mystery, illegibility, and inscrutability. (One might add to this list: unmappability.) In *Moby-Dick*, Ahab had asserted that what he hates more than anything else about the white whale is its inscrutability; Ahab too felt it as an oppressive closing in of the walls. "How can the prisoner reach outside except by thrusting through the wall? To me, the white whale is that wall, shoved near to me."[33] From the walls of a downtown Manhattan law office to the walls of the Tombs, where the "Egyptian character of the masonry weighed upon me with its gloom,"[34] the blankness and meaningless figured in the ambiguous urban space thwarts the representational project of literary narrative. In the global city, the cartographic project of *Moby-Dick* is suspended, as the "story of Wall Street" cannot read the text it places before us. In *Moby-Dick*, Melville could project a world that could, if only provisionally, map the world system in its fitful emergence and baroque complexity. With "Bartleby," in the depths of the financial district and figured, through it, in a world system beyond representation, Melville confronts a space that seems unchartable, a blank space that resists inscription.

Conclusion: "Leviathan is not the biggest fish," or, The Cartography of the Kraken

In "The Line," Melville describes the make and usage of the whale-line, the rope attached to the harpoon and used to capture the whale. Typical of so much of *Moby-Dick*, this chapter combines the power of narrative and description, as well as conceptually linking matters directly affecting the plot to a broader, more global realm of ideas and experiences. While the explication of the role and dangers of "the line" is necessary for understanding Ahab's fate at the novel's end, the chapter also displays a technique used by Melville to project his literary cartography of the world system. The "magical, sometimes horrible whale-line" is described in detail, from its substance (originally "the best hemp" but now "Manilla rope"), its width and length ("two-thirds of an inch" and "over two hundred fathoms"), and its place on the whale-ship ("coiled away in the tub," with "both ends of the line exposed," and threaded through the oarlocks and attached to the harpoon's "short-warp," which had previously undergone "mystifications too tedious to detail"). From these descriptions, Melville concludes:

> Thus the whale-line folds the whole boat in its complicated coils, twisting and writhing around it in almost every direction. All the oarsmen are involved in its perilous contortions; so that to the timid eye of the landsman, they seem as Indian jugglers, with the deadliest snakes sportively festooning their limbs. [. . .] For, when the line is darting out, to be seated then in the boat, is like being seated in the midst of the manifold whizzings of a steam-engine in full play, when every flying beam, and shaft, and wheel, is grazing you. It is worse; for you cannot sit motionless in the heart of these perils, because the boat is rocking like a cradle, and you are pitched one way or another, without the slightest warning; and only by a certain self-adjusting buoyancy and simultaneousness of volition and action, can you escape [. . .]

Melville ends this passage with the characteristic totalization of the partial experience, noting that "All men live enveloped in whale-lines. All are born with halters around their necks."[1]

The movement in this chapter from detailed description of a part of the whale-ship to the general connection to the whole and the ultimate fate of "all men" dramatizes Melville method in much of *Moby-Dick*. Jonathan Arac refers to this as "metonymizing": Beginning with "the ship's totality," Melville "metonymizes it into its parts," with chapter after chapter devoted to a distinct part of the ship ("The Mast-Head," "The Crotch," "The Forge"), the whale ("The Tail," "The Fountain," "The Cassock"), or the whaling industry ("The Dart," "The Specksynder," "The Gam"). "But from this partialization a new whole emerges."[2] Melville takes the discrete elements of the whale and whaling, focuses attention on the details while projecting a totality from them, and establishes a provisional representation of a world in the process. In writing this way, Melville is creating his own "lines," enveloping the reader in them, and forcing the reader to contemplate the world typified by such vastness and interconnectivity at once. As noted in Chapter 7, in his turn from the "plodding" personal narrative form to the baroque literary cartography, Melville longed to "plume [his] powers for a flight." As Gilles Deleuze has asserted, "To fly is to trace a line, lines, a whole cartography."[3] With *Moby-Dick*, he achieves such flight, drawing the lines of a postnational world system into a global map.

Melville's baroque mapping project is aimed at representing a seeming unrepresentable totality, figured forth in the whale and whaling, but embodied in the emergent postnational world system that has become our own dominant condition in the era of globalization. Whatever else the term might mean, and however one elects to value the phenomenon or condition, globalization has altered the fundamental structures of experience today. It is not simply that the age of market capitalism has given way to the age of imperialism or the age of multinational capitalism. More, it is a completely different understanding—or lack of understanding—of how relationships among the individual subject, society, and the world function. This is not to say that nationality is meaningless or disappears entirely; indeed, a certain resurgence of nationalisms seems to be a characteristic of the postmodern condition. But the nation-state is no longer the principal organizational form of power and thought in the era of globalization. The national study of literature, of course, is also somewhat anachronistic in such an era. The project of American Studies had been to register the distinctly unique nationality of writings produced in the United States, and *Moby-Dick* (among other works) became emblematic of such Americanness. But in our "borderless world,"[4] what is the value of such an emblem? To adapt a phrase of Fredric Jameson's, what is the role of the great American novel in the era of multinational capitalism?

Literary and cultural studies have begun to register this in the increasing transnationality of human experience, and a great deal of recent work may be evidence of a trans- or postnational world literature.[5] In such a postnational world system, what is the use of nation-specific studies at all? The time has come for a post-American Studies, with the concomitant scholarly activities of historical research into what "was" America and a speculative criticism of the world in which texts, reader, and writers inhabit. By my reading, *Moby-Dick* constitutes an attempt to map this world system, to make sense of the postnational reality. In a sense, then, *Moby-Dick* could be thought of as *postmodern* (as William Spanos does view it), though I am less interested in its postmodernity *per se* than in its ability to provide another occasion for us to think of our own postmodern condition. Our present may be likened to the baroque era, inasmuch as new, transnational organizations of power are compelling us to rethink the conceptual frameworks that have been in place for some time. A key element of this framework is the determining role of the nation-state and of national myths in the "life of the mind," in literature and cultural studies. Nearly two centuries after Goethe states that national literature has nothing much left to say, 150 years after Marx asserts that national and local literatures must give way to a *Weltliteratur*, and some fifty years after Erich Auerbach proclaims that "our philological home is the earth, it can no longer be the nation,"[6] the hypercanonization of *Moby-Dick* has kept the novel situated within a national tradition of American Studies. But the novel's postnationality makes it a particularly relevant text for our time, allowing us to think the system that so conditions our daily life, often in hitherto undreamed of ways.

Despite its actual decline around the time Melville was writing, the whaling industry itself makes for a proleptic figure of the postmodern or multinational industry of the late twentieth and early twenty-first centuries. A predecessor of the oil industry, which may be the very model of postmodern corporate subjectivity, whaling is among the first truly multinational businesses. Take the multinational labor force; it is not only that the crew is made up of individuals from all regions of the globe, but that, in C. L. R. James's words, "They are a world federation of modern industrial workers. They owe allegiance to no nationality."[7] Melville writes that "not one in two of the many thousand men before the mast in the American whale fishery, are American born," and this rough estimate seems to be borne out in the historical record.[8] An "Anacharsis Clootz deputation from all isles of the sea, and all ends of the earth," the crew members are themselves emblematic of a global industry. Certainly Melville could have populated the *Pequod* with Americans, even with people from the various regions of the United States, but Melville insists the multinational character of the industry, and hence of his own literary project.

The industry also operates in a way quite distinct from traditional national industries. Melville's whale fishery is an industry that involves a sort of "flexible production." In fact, it entails the full circuit of capitalist production—production, distribution, consumption—as well as the related processes of extracting raw materials, refining and packaging, storing and transporting the products, and so on. Much is made of the ship's transformation into a factory in "The Try-Works," but factory-production is but one aspect of the industry as Melville describes it. Whale oil is both the product of and the fuel for its own production (as "The Try-Works" and "The Lamp" emphasize). The actual work of producing it combines a primitive or primordial hunting with industrial production and even provides glimpses of a consumer society in which the consumers of the commodities are so distant, conceptually and geographical, from the production that they can have no real idea of it "truth." As Jameson has said of the existential condition of those in the metropole in the age of imperialism and monopoly capitalism, "the truth of that experience no longer coincides with the place in which it takes place."[9] As Melville repeatedly reminds us, the lights and corsets of advanced civilization rely on the far-flung voyages of the common whaleman. The whaling industry itself becomes a figure for the world system.

In addition to its dramatis personae and its elaboration of a global industry, *Moby-Dick* attempts to capture this world system by means of the novel's form. *Moby-Dick*'s "outrageous comprehensiveness of sweep," not only embraces a global geography but, as I have argued, also attempts to represent the globalization of discourse. With its inclusion of scientific writing, art history and criticism, poetry and a self-contained short story, philosophical reflection aplenty, practical descriptions of boats and boating, dirty jokes and lofty sentiment, so multiformal is this particular form-giving form that it is not always recognizable as a novel. This is, in part, what Franco Moretti means when he writes that the modern epic, the "world text," is the symbolic form of the world system (and that the novel is the proper symbolic form of the nation-state). Melville, in *Moby-Dick,* wishes to capture the image of the entire world by including in its form nearly everything in the entire world. It is not only that all of these things, and more, are included, but also how they are organized. The literary overview enables the novel to map an ever-more-complicated social space that the individual subject can no longer adequately or confidently maneuver within. Combined with *Moby-Dick*'s panoramic overview, and enabling its functions, is the novel's delving into nit-picky detail, dwelling at length on mere parts of the ship or of the whale; this "minutest detail" grounds the novel in a mock-scientific practice, the proper sphere of the conceptual personae (such as the "pale Usher" whose concentration is focused on a single word, and the "poor devil of a Sub-Sub," who is a "mere painstaking burrower and grubworm," along with the other seekers and custodians of knowledge along the way). The movements from the microscopic to the macrocosmic, more than the actual plot, define the rhythm of the novel, and suggest a view of the world. With painstaking detail combined with vast spaces, Melville's produces a map of the world system.

Like the line that envelops the whalemen, and all humanity, the transnational interconnectivity of globalization has bound people and places together in ways only remotely dreamed of in the nineteenth century. It goes beyond mere advancements in communications technology or international capitalism, effecting a real change in the way even the details of one's personal life are experienced. As the most recent global financial crisis has brought home to many, a system beyond any form of regional or national regulation is now in place, but is also unlocalizable in a particular place, that is difficult for even those who are most intimately involved in it to understand. For the rest of us, merely trying to get a sense of our place in a global system in which one's entire life's savings—or even an entire national treasury—can be inadvertently wiped out by a keystroke in New York or London or Hong Kong. In the interconnected and baroque world of global finance, for example, the very system that conditions the minutest aspects of our lives may be completing unknown or unknowable. "How does one know about, or demonstrate against, an unlisted, virtual, offshore corporation that operates in an unregulated electronic space using a secret proprietary trading strategy to buy and sell arcane financial instruments?"[10] The representational crisis so often associated with the onset of modernity, and intensified along the moments of further modernization and postmodernization, is all the more critically urgent in a world system in which the traditional social and spatial landmarks are unavailable. The literary cartography of this world system is both more urgent and more difficult than ever.

As Melville discovered—first in his "naïve" personal narratives of the unrepresentably vast and varied space of the Pacific, then in his grandly baroque exposition and projection of a world in *Moby-Dick*, and with his failure to capture the ambiguous urban space, and his ultimate capitulation to the inscrutability and unknowability of these spaces—the mapping project rarely if ever succeeds, and then, only provisionally. The cartographic anxiety that one might feel when faced with any spatially disorienting condition is all the more compounded in attempting to map a totality. Yet, in order to orient oneself, one has to map this space, however impossible the project may be. This is a fundamentally *utopian* project. Of course, the map, as always is utopian; there can be no true maps. Literally utopian, from a certain point a view, the map depicts a *no-place*, with respect to both the no-wheres figured on the map itself and the imaginary-yet-real system that cannot be pinned down on a grid or spatial array. Melville's project is utopian in another sense: utopia is not merely the imaginary depiction of a nonexistent, and perhaps ideal, *topos*; utopian discourse has also always maintained a critical function. In imagining a radical alternative to the actually existing place, utopia establishes a critique of the society itself (as is clearly visible in Thomas More's own *Utopia* or, say, George Orwell's apparent dystopias). Utopian discourse emerges in a transitional moment, in what Louis Marin (in *Utopics*) calls "the neutral": *ne-uter*, neither one nor the other. In More's case, he was writing from a historical position in which feudalism is dying out but also in which capitalism is not yet widespread as a distinct mode of production; and politically, this is a moment of decay for the system of disparate estates at the same time as the absolute state had not really emerged as the dominant form of social organization. As Marin argues, this neutrality characterizes the distinct nature of utopian writing as a genre. Melville, in his utopian mapping of an emergent world system, criticizes the dominant national narrative of his own day and proleptically criticizes the project of twentieth-century American studies. As I have argued all along, Melville does not fit in this tradition. Standing between the dominant nationalism of his day and the emergent postnationalism, visible in its barest outlines in the whaling industry and the development of postnational forces in the cities and the edges of the world, Melville projects a no-place and every-place, a world system larger and more complex than the Hobbesian leviathan of his day.

If, as Jameson asserted some time ago, the "new world system" is "the true ground of Being" in our time, and the political form of postmodernism must be "the invention and projection of a global cognitive mapping, on a social as well as a spatial scale,"[11] then Melville at least provides some clues—bread crumbs left on the path of his own transitional moment for us to follow now—as to the nature and practice of the work to be done. In a famous letter to Hawthorne, written not long after he had published *Moby-Dick,* Melville avers, "Leviathan is not the biggest fish; — I have heard of Krakens."[12] The urgent project of criticism and literature in our own time, in the postmodern condition of the era of globalization, may find its vocation in the cartography of the kraken.

Notes

Preface: "When Leviathan's the text"

[1] Herman Melville, *Moby-Dick, or, The Whale*, ed. Harrison Hayford et al. (Evanston and Chicago: Northwestern University Press and the Newberry Library, 1988), 71–72.

[2] "In its vast spaces and in Melville's blazingly original style, *Moby-Dick* is about (it seems silly to say it this way) the whole world; it willingly incorporates everything, leaving such small matters as resolution, inconsistency, and indeed evaluating the *consequences* of so tremendous as shattering an experience to lesser natures." Edward Said, "Introduction to *Moby-Dick*," in *Reflections on Exile* (Cambridge: Harvard University Press, 2000), 369.

[3] *Moby-Dick*, 456.

[4] Herman Melville, *Mardi, and A Voyage Thither*, ed. Harrison Hayford et al. (Evanston and Chicago: Northwestern University Press and the Newberry Library, 1970), 556.

[5] *Moby-Dick*, 198.

[6] Peter Turchi, *Maps of the Imagination: The Writer as Cartographer* (San Antonio: Trinity University Press, 2004).

[7] See Fredric Jameson, *The Political Unconscious: Narrative as a Socially Symbolic Act* (Ithaca: Cornell University Press, 1981), 102; see also his *The Geopolitical Aesthetic: Cinema and Space in the World System* (Bloomington: Indiana University Press, 1993), 82.

[8] See Donald E. Pease, "National Identities, Postmodern Artifacts, and Postnational Narratives," in *National Identities and Post-Americanist Narratives*, ed. Donald E. Pease (Durham: Duke University Press, 1994), 3–4; Jonathan Arac, "*Huckleberry Finn*" *as Idol and Target: The Function of Criticism in Our Time* (Madison: University of Wisconsin Press, 1997), 137–38.

[9] *Moby-Dick*, 455.

[10] *Moby-Dick*, 455, 134.

Chapter 1

[1] For a useful summary on the relationship between the canonization of Melville and the establishment of a canon of American literature in general, see William V. Spanos, *The Errant Art of 'Moby-Dick': The Canon, the Cold War, and the Struggle for American Studies* (Durham: Duke University Press, 1995), especially 1–42.

[2] Gilles Deleuze and Claire Parnet, *Dialogues*, trans. Hugh Tomlinson and Barbara Habberjam (New York: Columbia University Press, 1987), 15.

[3] The term *postnational* here refers to the condition in which the territorial nation-state is no longer the locus classicus of culture, economics, or even politics. I specifically refer to the era of globalization or multinational capitalism, in which the old categories no longer carry the same explanatory power. See Jürgen Habermas, *The Postnational Constellation*, trans. Max Pensky (Cambridge: MIT Press, 2001). Donald Pease has employed the term to refer to those forces—including those of "previously subjugated social categories of race, class, and gender"—that disrupt the social categories of national narratives, without particular reference to globalization, but with reference to postmodernism, which may be thought of as the "cultural logic" of the age (á la Fredric Jameson). See Pease, "National Identities, Postmodern Artifacts, and Postnational Narratives," in *National Identities and Post-Americanist Narratives*, ed. Donald E. Pease (Durham: Duke University Press, 1994), 4–5. I directly associate the *postnational* condition with the world system in the era of globalization.

[4] The term "hypercanonization" is used by Jonathan Arac to designate the critical and curricular dominance of a few works in American literary studies of a few works—in fiction, pre-eminently *The Scarlet Letter, Moby-Dick,* and *Huckleberry Finn.* In Arac's view, hypercanonization leads to a public culture of idolatry, where the text is viewed as above critical reproach. See Jonathan Arac, *Huckleberry Finn as Idol and Target: The Functions of Criticism in Our Time* (Madison: University of Wisconsin Press, 1997). For another view of *Moby-Dick* as a "sacred text," see Franco Moretti, *Modern Epic: The World System from Goethe to García Márquez*, trans. Quentin Hoare (London: Verso, 1994).

[5] For Deleuze's argument that Spinoza's scholia form a second, more practical *Ethics* running parallel to the more rational *Ethics* of the propositions and proofs, see his *Expressionism in Philosophy: Spinoza*, trans. Martin Joughin (New York: Zone Books, 1990), 337–350. Deleuze expressly calls the former a "subterranean *Ethics*" in *Spinoza: Practical Philosophy*, trans. Robert Hurley (San Francisco: City Lights, 1988), 29.

[6] C. L. R. James, Mariners, Renegades, and Castaways: The Story of Herman Melville and the World We Live In (Hanover, New Hampshire: University Press of New England, 2001), 3.

[7] My use of the terms *emergent, dominant,* and *residual* follows that of Raymond Williams, *Marxism and Literature* (Oxford: Oxford University Press, 1977), 121–127.

[8] Fredric Jameson, *The Geopolitical Aesthetic: Space and Cinema in the World System* (Bloomington: Indiana University Press, 1992), 15.

[9] Fredric Jameson, *Postmodernism, or, The Cultural Logic of Late Capitalism* (Durham: Duke University Press, 1991), 418.

[10] Indeed, the title was not Matthiessen's, but rather was offered to him by a younger colleague. In his autobiographical preface to *The Power of Blackness*, Harry Levin notes "by ever-grateful acknowledgment" that he had in fact suggested the title *American Renaissance* to Matthiessen, who had wanted to call the book *Man in the Open Air* ("after an apt phrase from Whitman"), but the "publisher

[had] wanted something more descriptively categorical." See Harry Levin, *The Power of Blackness* (Athens, Ohio: Ohio University Press, 1989), vii.

[11] F. O. Matthiessen. American Renaissance: Art and Expression in the Age of Emerson and Whitman (Oxford: Oxford University Press, 1941), vii.

[12] Jonathan Arac, "F. O. Matthiessen: Authorizing the American Renaissance," in *The American Renaissance Reconsidered*, eds Walter Benn Michaels and Donald E. Pease (Baltimore: Johns Hopkins University Press, 1985), 94.

[13] See my "'Believing in America': The Politics of American Studies in a Post-national Era." *The Americanist* XXIII (2006): 69–81.

[14] René Wellek, *Concepts in Criticism* (New Haven: Yale University Press, 1963), 116.

[15] Heinrich Wölfflin, *Renaissance and Baroque*, trans. Kathrin Simon (Ithaca: Cornell University Press, 1964), 15.

[16] Ibid., 23.

[17] Ibid., 23.

[18] Antonio Negri, *The Political Descartes*, trans. Matteo Mandarini and Alberto Toscano (London: Verso, 2007), 114–15.

[19] Mikhail Bakhtin, "Discourse in the Novel," in *The Dialogic Imagination*, trans. Caryl Emerson and Michael Holquist (Austin: University of Texas Press, 1981), 394.

[20] Walter Benjamin, *The Origin of German Tragic Drama*, trans. John Osborne (London: Verso, 1977), 55.

[21] Ibid., 195.

[22] Jameson, *Marxism and Form* (Princeton: Princeton University Press, 1971), 72.

[23] John Beverley, "On the Spanish Literary Baroque," in *Against Literature* (Minneapolis: University of Minnesota Press, 1993), 48.

[24] Karl Marx and Friedrich Engels, *The Communist Manifesto* (New York: Signet, 1998), 54.

[25] Evert Duyckinck, "Melville's Moby-Dick; or, the Whale," *The Literary World* (November 22, 1851): 403–04.

[26] Henri Focillon, *The Life of Forms in Art*, trans. Charles B. Hogan and George Kubler (New York: Zone Books, 1992), 58.

[27] "Sir Nathaniel" [but possibly William Harrison Ainsworth], "American Authorship No. 4—Herman Melville," New Monthly Magazine [London] (July 1853): 307–08. Reprinted in Jay Leyda, *A Melville Log*, vol. 1 (New York: Harcourt, 1951), 477. Emphasis added.

[28] What follows in this section is based in large part on the Immanuel Wallerstein's influential history of the modern world system in the sixteenth and seventeenth centuries. See *The Modern World-System*, 3 vols. (New York: Academic Press, 1974).

[29] Gilles Deleuze, *The Fold: Leibniz and the Baroque*, trans. Tom Conley (Minneapolis: University of Minnesota Press, 1993), 68.

[30] Wallerstein, *The Modern World-System*, vol. II, 7.

[31] Michel Foucault, "Nietzsche, Genealogy, History," in *Language, Counter-Memory, Practice*, trans. Donald F. Bouchard and Sherry Simon (Ithaca: Cornell University Press, 1977), 148, 149–150.

[32] Carl J. Friedrich, *The Age of the Baroque, 1610–1660* (New York: Harper & Row, 1952), 1.

[33] Wallerstein, *The Modern World-System*, vol. I, 145.

[34] See, e.g. Raymond Williams, *Keywords* (Oxford: Oxford University Press, 1976), 213–214.

[35] Benedict Anderson, *Imagined Communities: Reflections on the Origin and Spread of Nationalism* (London: Verso, 1991). Arguably, this has changed in recent years, suggesting that the dominance of the national and the nation-state form is at an end. In the immediate post-Cold War years, a resurgence of what might be called sub- or intra-national ethnic identities in Eastern Europe and elsewhere led to conflict precisely because the national borders did not contain the "nation" imagined. Newly recognized but often pre-modern "nations" (in that pre-modern sense), were disjoined from the states to which they were hitherto attached or sought new nation-states of their own. At the same time, greater "supra-national" identities—such as the transnational Islam, for instance—also arose. For such groups a nation-state cannot and should not delimit the parameters of the group identity. In ironic coincidence with these movements, postnational or multinational capitalism has in fact broken the economic ties between a "people" and the territorial state, making the world more postnational even as products may be marketed to ever-more-discretely identifiable groups.

[36] Arac, *The Emergence of American Literary Narrative* (Cambridge: Harvard University Press, 2005), 3.

[37] Wallerstein, *Modern World-System*, vol. I, 349.

[38] Indeed, one might say that the nineteenth-century United States contained something like Wallerstein's "world-economy" in microcosm, with the industrial and mercantile north forming a core, the agrarian and slaveholding south as the periphery, and the newly settled west (actually, what is today considered midwest) and Appalachian regions semiperipheral.

[39] For a number of differing views on this subject, see Franco Moretti, ed., *The Novel*, 2 vols. (Princeton: Princeton University Press, 2006).

[40] Georg Lukács, *The Theory of the Novel*, trans. Anna Bostock (Cambridge: The MIT Press, 1971), 29.

[41] Martin Heidegger, *Being and Time*, trans. John Macquarrie and Edward Robinson (New York: Harper and Row, 1962), 233.

[42] Lukács, *Theory of the Novel*, 60.

[43] Michel Foucault, *The Order of Things*, trans. anon. collective (New York: Vintage, 1973), 48.

[44] Ibid., 47–48.

[45] Lukács, Theory of the Novel, 88.

[46] Ibid., 101.

[47] Mikhail Bakhtin, "Epic and Novel," in *The Dialogic Imagination*, trans. Caryl Emerson and Michael Holquist (Austin: University of Texas Press, 1981), 15.

[48] See Jameson, *Postmodernism*, 51–52.

[49] Jorge Luis Borges, "Of Exactitude in Science," in *A Universal History of Infamy*, trans. Norman Thomas de Giovanni (London: Penguin, 1975).

[50] Jameson, *The Political Unconscious: Narrative as a Socially Symbolic Act* (Ithaca: Cornell University Press, 1981), 104.

[51] Thomas Pynchon, *The Crying of Lot 49* (New York: Harper and Row, 1990), 81–82.

Chapter 2

[1] Charles Olson, *Call Me Ishmael* (San Francisco: City Lights, 1947), 11.
[2] Nathaniel Hawthorne, *The House of the Seven Gables* (Oxford: Oxford University Press, 1991), 33.
[3] Ibid., 19.
[4] Marx, *The Eighteenth Brumaire* (New York: International Publishers, 1963), 15.
[5] Hawthorne, *House*, 297.
[6] Throughout the novel, Hawthorne offers mysterious phenomena only to counter-offer a rather banal, rational explanation. Specific examples in the text may be too numerous to mention, from the brackish water in Maule's cursed well that might been dug too deeply to Hepzibah's ill-tempered scowl being a mere side-effect of her near-sightedness. Hawthorne is careful (as he states in the preface) "to mingle the Marvellous rather as a slight, delicate, and evanescent flavor, than as any portion of the actual substance of the dish offered to the Public" (1).
[7] Jonathan Arac, *Commissioned Spirits* (New York: Columbia University Press, 1989), 3.
[8] Hawthorne, *House*, 3.
[9] D.H. Lawrence, *Studies in Classic American Literature* (New York: Viking, 1964), 5–6.
[10] This work was originally published as "Narrative Forms," in *The Cambridge History of American Literature, Vol. 2: 1820–1860*, ed. Sacvan Bercovitch (Cambridge: Cambridge University Press, 1995).
[11] Jameson, The Geopolitical Aesthetic, 2.
[12] Arac, *Emergence*, 1–2.
[13] See Terry Eagleton, *Literary Theory*, 2nd edn (Minneapolis: University of Minnesota Press, 1996), 15.
[14] Arac, *Emergence*, 3.
[15] The appropriation of these works by American Studies for use in a larger national narrative—e.g. in which Ishmael can be viewed as a representative democrat struggling against Ahab's totalitarianism—represent what Arac calls "the nationalization of literary narrative." See his *Huckleberry Finn as Idol and Target* (Madison: University of Wisconsin Press, 1997), especially 133–39.
[16] Donald Pease, "National Identities," 4.
[17] See Leo Marx, "On Recovering the 'Ur' Theory of American Studies." *American Literary History*, 17.1 (Spring 2005), 118–134.
[18] Sacvan Bercovitch, *The Rites of Assent* (London: Routledge, 1993), 358.
[19] James Fenimore Cooper, *The Last of the Mohicans* (New York: Penguin, 1986), 7.
[20] For example, the third chapter of Bancroft's first volume, which deals with the period from Columbus to 1660, is entitled "England Takes Possession of the United States." See Arac, *Emergence*, 27.
[21] Bercovitch, *The Rites of Assent*, 6–7.

[22] Washington Irving, "The Legend of Sleepy Hollow," in *The Sketch Book* (New York: Penguin, 1961), 331.

[23] Longstreet first published his work in 1835 anonymously, as Georgia Scenes, Characters, Incidents, &c. in the first Half Century of the Republic: By a Native Georgian. See Arac, *Emergence*, 37.

[24] Davy Crockett is a case in point. Initially a unique figure, who moreover was deployed *in opposition to* a perceived national culture by Whigs who denounced Andrew Jackson's regime, Crockett moves from the local narrative of a backwoods braggart who was decidedly outside of the mainstream national culture, to a representative national character himself. As Arac notes, "By the time Walt Disney took them up into massively circulated television, film, and song materials in the 1950s, they [i.e. the Crockett narratives] had become integrated into national narrative, standing no longer for a part or a party but for the whole of America." Arac, *Emergence*, 35.

[25] Hawthorne, "Ethan Brand," in *Young Goodman Brown and Other Tales*, ed. Brian Harding (Oxford: Oxford University Press, 1987), 329–30.

[26] Arac, *Emergence*, 61–62.

[27] Bloom, "Introduction," in *Bloom's Classic Critical Views: Edgar Allan Poe*, ed. Robert T. Tally Jr. (New York: Chelsea House, 2007), xi.

[28] Matthiessen, *American Renaissance* (Oxford: Oxford University Press, 1941), xii, n.3.

[29] Arac, *Emergence*, 68.

[30] Ibid., 76.

[31] Arac, *Emergence*, 77.

[32] See Jay Leyda, *A Melville Log*, 196.

[33] Herman Melville, *Typee*, ed. Harrison Hayford et al. (Evanston and Chicago: Northwestern University Press and the Newberry Library, 1968), xiv.

[34] Hawthorne, *The Blithedale Romance* (Oxford: Oxford University Press, 1991), 2.

[35] See Arac, "The Politics of *The Scarlet Letter*," in *Ideology and Classic American Literature*, ed. Sacvan Bercovitch and Myra Jehlen (Cambridge: Cambridge University Press, 1986), 247–66.

[36] Poe, "Twice-told Tales," in *The Fall of the House of Usher and Other Writings* (New York: Penguin, 1986), 387–88.

[37] Hawthorne, *Letters, 1853–1856* (Columbus: The Ohio State University Press, 1987), 304.

[38] Melville, *Correspondence*, eds, Harrison Hayford, Alma A. MacDougall, G. Thomas Tanselle et al. (Evanston and Chicago: Northwestern University Press and the Newberry Library, 1987), 192.

[39] Hawthorne, The House of the Seven Gables, 1.

[40] Melville, *Moby-Dick*, 6.

[41] Arac, *Emergence*, 241.

[42] For a powerful, and now canonical, treatment of these two aspects of space and how they interact and merge, see Henri Lefebvre, *The Production of Space*, trans. Donald Nicholson-Smith (Oxford: Blackwell, 1991).

[43] Wai Chee Dimock has recently called for the alterative history and geography of a "planetary" perspective in reshaping American studies. See Dimock and

Lawrence Buell, eds, *Shades of the Planet: American Literature as World Literature* (Princeton: Princeton University Press, 2008).

Chapter 3

1 Melville, *Redburn: His First Voyage*, eds Harrison Hayford, Hershel Parker, and G. Thomas Tanselle (Evanston and Chicago: Northwestern University Press and the Newberry Library, 1969), 165.
2 Melville, *Moby-Dick, or, The Whale*, eds Harrison Hayford, Hershel Parker, and G. Thomas Tanselle (Evanston and Chicago: Northwestern University Press and the Newberry Library, 1988), 121.
3 *Redburn*, 169.
4 Spengemann, "Introduction," to Melville, *Pierre* (New York: Penguin, 1996), ix.
5 Benedict Anderson's argument on the relationship between the emergence of the novel and the formation of nationalisms remains the canonical treatment. See *Imaginary Communities* (London: Verso, 1991), 24–25.
6 See Franco Moretti, *Modern Epic: The World System from Goethe to García Márquez*, trans. Quentin Hoare (London: Verso, 1994).
7 *Moby-Dick*, 273.
8 Goethe, "On World Literature," *Essays on Art and Literature*, ed. John Geary, trans. Ellen von Nardhoff and Ernest H. von Nardhoff (Princeton: Princeton University Press, 1994), 224–228.
9 Marx and Engels, *The Communist Manifesto*, trans. anonymous (New York: Signet, 1998), 54–55.
10 Marx, "Bastiat and Carey," *Collected Works of Karl Marx and Friedrich Engels*, vol. 28, translated by Ernst Wangermann (Moscow: International Publishers, 1986), 6.
11 Ibid., 7.
12 Ibid., 8–9.
13 Isaiah Berlin, *The Crooked Timber of Humanity* (Princeton: Princeton University Press, 1998), 58.
14 Arac, *The Emergence of American Literary Narrative* (Cambridge: Harvard University Press, 2005), 241.
15 See Poe, "Twice-told Tales," in *The Fall of the House of Usher and Other Writings*, ed. David Galloway (New York: Penguin, 1986), 395. It should be noted that Poe also disputes that a novel can present the true sense of totality, since it requires too much time to read, and during the intervening (and interrupting) time, the reader will lose the totality or the singularity of effect.
16 Melville, *Moby-Dick*, xv–xvii.
17 Arac, *Emergence*, 167.
18 For a fascinating perspective on "international literary space" and world literature, see Pascale Casanova, *The World Republic of Letters*, trans. M. B. DeBevoise (Cambridge: Harvard University Press, 2004).
19 Evert Duyckinck, "Melville's Moby-Dick; or, the Whale," *The Literary World* (November 22, 1851): 403.

[20] Moretti, *Modern Epic*, 50.

[21] Ibid., 4, note 2.

[22] Melville, *Moby-Dick*, 145.

[23] *Modern Epic*, 51.

[24] Moretti notes that his book was originally intended as a study of modernism, so extending his temporal frame from the time of the First World War all the way back to 1800 was already a stretch.

[25] Jameson, *The Geopolitical Aesthetic: Cinema and Space in the World System* (Bloomington: Indiana University Press, 1992), 82.

Chapter 4

[1] In the first-published English edition (bearing the title, *The Whale*), these chapters appear at the end of the novel as an appendix. That edition also famously omitted the Epilogue, leading to confusion among many reviewers. The first American edition properly placed "Etymology" and "Extracts" in the fore, with small roman page numbers. See Herman Melville, *Moby-Dick, or, The Whale*, eds Harrison Hayford, Hershel Parker, and G. Thomas Tanselle (Evanston and Chicago: Northwestern University Press and the Newberry Library, 1988).

[2] The pervasiveness of the interpretations of the novel which follow from the "Call me Ishmael" beginning may be illustrated by comparing two important and quite different Melville scholars. Howard P. Vincent, whose *The Trying-out of Moby-Dick* was among the first comprehensive readings of the novel, utterly ignores "Etymology" and "Extracts." Vincent's interpretation begins (in a chapter significantly titled "Narrative Beginnings"): " 'Call me Ishmael.' Arrestingly, dramatically, as though a mysterious dissonance had been plucked by the massed strings, Melville's great whaling symphony begins." *The Trying-out of "Moby-Dick"* (Boston: Houghton Mifflin, 1949), 55. Similarly, William V. Spanos, in *The Errant Art of "Moby-Dick"*—a book expressly opposed to the sort of "Old Americanist" approach exemplified by Vincent's study (among many others)—examines in detail the minutiae as well as the grand expositions of the novel, but he too manages to overlook entirely the unnumbered chapters preceding "Loomings," choosing instead to open his interpretation with the words: " 'Call me Ishmael.' So, unexpectedly, Melville begins *Moby-Dick*." *Errant Art* (Durham: Duke University Press, 1995), 75. Both Vincent and Spanos, notably opposed on many other critical matters, isolate the famous three-word sentence *as* the beginning; both use the verb *to begin*. I mention these two works and these two scholars, in particular, out of the hundreds of studies of *Moby-Dick*, because few others have attempted such a comprehensive and thoroughgoing analysis of the novel; their comprehensiveness makes their omissions all the more striking.

[3] As used here, *national narrative* incorporates the nineteenth-century narrative form (discussed in Chapter 2 above) and the broader mythic narrative embrace in twentieth-century American Studies. In this chapter, its force derives primarily from what Arac calls the "nationalization of literary narrative" in the twentieth century. That is, with the canonization of *Moby-Dick*, a generation of

critics established an approach to the novel that made it fit within a discourse of national narrative, even as they idolized the novel for its inherently literary and aesthetic qualities. By *postnational* in this context, I mean an approach that no longer views the nation or nationalism as the central, organizing concept of literary or cultural studies; more positively, I mean the recognition of a post- or supranational position that may encompass a "world" view, beyond any national boundaries. See, e.g. Wai Chee Dimock, *Through Other Continents: American Literature Across Deep Time* (Princeton: Princeton University Press, 2006).

4 See Dimock, *Empire for Liberty* (Princeton: Princeton University Press, 1989), 25–26, 38–41. Dimock's reading, in her own words, "virtually ignores" Ishmael in name, and Dimock does not see in Ishmael a figure of redemption (see 236 n.52). However, in placing the blame for "blaming the victim" squarely at Melville's own feet, Dimock identifies the "whale author" of "an imperial folio" as the representative American subject himself: "In any case, as the exemplar of a 'poetics of individualism,' Melville will emerge in my account as something of a representative author, a man who speaks for and with his contemporaries, speaking for them and with them, most of all, when he imagines himself to be above them" (6). Dimock thus turns Melville into the very thing that Ishmael had been to other critics, the avatar of American national culture.

5 Pease, "National Identities, Postmodern Artifacts, and Postnational Narratives," in *National Identities and Post-Americanist Narratives*, ed. Donald E. Pease (Durham: Duke University Press, 1994), 3–4. These key terms will be recognized as the titles of three "classic" works in American Studies: R. W. B. Lewis's *The American Adam* (1955), Henry Nash Smith's *Virgin Land* (1950), and Perry Miller's *Errand into the Wilderness* (1956).

6 The development of this "Ishmael vs. Ahab" argument is discussed in more detail by Pease in his influential essay, "*Moby Dick* and the Cold War," in *The American Renaissance Reconsidered*, eds Walter Benn Michaels and Donald E. Pease (Baltimore: Johns Hopkins University Press, 1985), 113–55.

7 Evert Duyckinck, "Melville's Moby-Dick; or, the Whale," *The Literary World* (November 22, 1851): 403–04.

8 "Sir Nathaniel" [but possibly William Harrison Ainsworth], "American Authorship No. 4—Herman Melville," New Monthly Magazine [London] (July 1853): 307–08. Reprinted in Jay Leyda, *A Melville Log*, vol.1 (New York: Harcourt, 1951), 477. Indeed, another anonymous reviewer had asserted that Ahab's "ravings, and the ravings of some of the tributary characters, and the ravings of Mr. Melville himself, meant for eloquent declamation, are such as would justify a *writ de lunatico* against all parties." See Leyda, 446.

9 Anon., *United States Magazine and Democratic Review* (January 1852): 93. Reprinted in Leyda, 445–46.

10 The quotes are from reviews in the *Athenaeum* (October 25, 1851) and the *Literary Gazette* (December 6, 1851), and are reprinted in Leyda, 430–31 and 439.

11 Spanos, *The Errant Art of "Moby-Dick"*, 75.

12 For Spanos's analysis of Father Mapple's sermon, see *Errant Art*, especially 87–104.

13 Manifest Destiny, as an idea and as an ideology, is the very figure of the movement from the *theo*-logocentrism of the Puritan errand into the wilderness to the *anthropo*-logocentrism of nineteenth-century America.

[14] Spanos, *Errant Art*, 75.
[15] Franco Moretti, *Modern Epic* (London: Verso, 1996), 62–63.
[16] Spanos, *Errant Art*, 192.
[17] Edward Said, *Beginnings* (New York: Columbia University Press, 1985), 5.
[18] Erich Auerbach, "Philology and *Weltliteratur*," trans M. and E. W. Said, *Centennial Review* 13 (Winter 1969): 15.
[19] Said, *Beginnings*, 72–73.
[20] Ibid., 73.
[21] Melville, *Moby-Dick*, xv.
[22] Ibid., 456.
[23] Ibid., 264, 118, 127–28, 379.
[24] Ihab Hassan specifically links this unrepresentability to a mapping project, what he calls a "poetics of cartography" in *Moby-Dick*. See his "Maps and Stories: A Brief Meditation," *Georgia Review* 59.4 (Winter 2005): 751–63.
[25] *Moby-Dick*, 123.
[26] For a discussion of the link between Ahab and the whale, and more generally between the human and natural world, see Armstrong, " 'Leviathan is a Skein of Networks,' " *ELH* 71 (2004): 1039–63.
[27] *Moby-Dick*, 194–95, 306.
[28] Ibid., 571.
[29] Ibid., xv, 572, emphasis added.
[30] In arguing in favor of this alternative line of thought in *Moby-Dick*, I am also arguing against the heightened significance attached to the Epilogue itself, which—operating like Freudian *Nachträglichkeit*—retroactively establishes, or re-establishes, the personal narrative of Ishmael *as* a survivor's tale.
[31] Comparing Melville's approach to history with Marx's, Loren Goldner has argued that Melville viewed historical time in "vorticist-helical" terms, with express reference to vortices and spirals. See his *Herman Melville: Between Charlemagne and the Antemosaic Cosmic Man* (New York: Queequeg Publications, 2007), especially 100–16.
[32] Yunte Huang has noted that the "careful disorderliness" of the list involves a kind of world history, although he views Ishmael *as* the historian. See his *Transpacific Imaginations: History, Literature, Counterpoetics* (Cambridge: Harvard University Press, 2008), 72–81.
[33] For a fascinating discussion of "Etymology," especially with respect to the potentially rich meanings associated with the "incorrectly written" Hebrew word for *whale*, see Neil Schleifer, "Melville as Lexicographer," *Melville Society Extracts* 98 (1994): 1–6.
[34] See the dialectic of Global and Babel in the postnational era, see Jonathan Arac, "Global or Babel: Language and Planet in American Literature," in *Shades of the Planet*, eds Wai Chee Dimock and Lawrence Buell (Princeton: Princeton University Press, 2007), 19–38.
[35] Bakhtin, "Discourse in the Novel," in The Dialogic Imagination, trans. Michael Holquist (Austin: University of Texas Press, 1981), 262.
[36] *Moby-Dick*, xvii, xix, xxiii.
[37] Ibid., xx.

[38] Another extract, attributed to Edmund Burke, mentions the whale only metaphorically and directly addresses the nation-state: "Spain—a great whale stranded upon the shores of Europe" (xxiii). Here again, the leviathan figure of the modern state is invoked (more noteworthy still since it is Spain, arguably the first and initially most powerful modern state), but it is invoked only to show its relative impotence in relation to the world system, here presented in the figure of Europe.

[39] On "conceptual personae," see Delueze and Guattari, *What is Philosophy?*, trans. Hugh Tomlinson and Grant Burchell (New York: Columbia University Press, 1994), 63.

[40] *Moby-Dick*, xv, xii.

[41] Claude Richard, *American Letters*, trans. Carol Mastrangelo Bové (Philadelphia: University of Pennsylvania Press, 1998), 60.

[42] *Moby-Dick*, 456, 451, 263.

[43] Ibid., 374.

[44] Spanos, *Errant Art*, 86.

[45] Moretti, *Modern Epic*, 50.

Chapter 5

[1] Emerson, "History," in *Selected Essays*, ed. Larzer Ziff (New York: Penguin, 1982), 161.

[2] Melville, *Moby-Dick, or, The Whale*, eds Harrison Hayford, Hershel Parker, and G. Thomas Tanselle (Evanston and Chicago: Northwestern University Press and the Newberry Library, 1988), 117, 121.

[3] Heimert, "*Moby-Dick* and American Political Symbolism." *American Quarterly* 15.4 (Winter 1963): 498–534.

[4] Deleuze, *Difference and Repetition*, trans. Paul Patton (New York: Columbia University Press, 1994), 36.

[5] Gilles Deleuze and Claire Parnet, *Dialogues*, trans. Hugh Tomlinson and Barbara Habberjam (New York: Columbia University Press, 1987), 14–15. The other quotation in this paragraph is from p. 13.

[6] Deleuze discusses Melville in *Dialogues* and *A Thousand Plateaus*, among other places, and he devoted a lengthy article to a novel interpretation of "Bartleby, the Scrivener." See "Bartleby, or, the Formula," in *Essays Critical and Clinical*, trans. Daniel W. Smith and Michael A. Greco (Minneapolis: University of Minnesota Press, 1997), 68–90.

[7] It is probably unlikely that Melville read Kant directly, although he mentions Kant in his works (as in the comparison between Locke and Kant in *Moby-Dick* ["Stubb and Flask Kill a Right Whale," 327]). Nancy Fredericks argues that "Melville probably never read Kant" (*Melville's Art of Democracy* [Athens: University of Georgia Press, 1995], 12), and Merton Sealts has suggested that Melville picked up his Kantian philosophy second-hand from Emerson (*Melville's Reading: A Checklist of Books Owned or Borrowed* [Madison: University of Wisconsin Press, 1966], 318). It is, however, well known and on Melville's own authority

that he discussed Kant with his friend and companion on a transatlantic voyage in October 1849, George Adler, a German philologist versed in Romantic literature and philosophy. In his journal, Melville wrote: "We had an extraordinary time & did not break up till after two in the morning. We talked metaphysics continually, & Hegel, Schlegel, Kant &c were discussed under the influence of the whiskey" (*Journals* 8).

[8] Kant, *Critique of Pure Reason*, trans. Norman Kemp Smith (London: Macmillan, 1933), 8–9.

[9] Ibid., 30.

[10] Ibid., 422–23.

[11] Ibid., 423–24. The references to "impartial umpires" and a "dialectical battlefield" may be found on pp. 394–95.

[12] Ibid., 424–25, 429.

[13] Donald E. Pease, "National Identities, Postmodern Artifacts, and Postnational Narratives," in *National Identities and Post-Americanist Narratives*, ed. Donald E. Pease (Durham: Duke University Press, 1994), 3–4.

[14] Notably, by Wai Chee Dimock in *Empire for Liberty: Melville and the Poetics of Individualism* (Princeton: Princeton University Press, 1989).

[15] *Moby-Dick*, 273.

[16] Ibid., 457.

[17] And, of course, *Adam* (literally and etymologically) is merely the name for *man.*

[18] *Moby-Dick*, 462.

[19] Ibid., 182–83.

[20] Although these images are crucial to the national narratives of Cooper and Bancroft, both writers were well aware of the importance of the sea.

[21] *Moby-Dick*, 273–74.

[22] Ibid., 572.

[23] Ibid., 274.

[24] Alexis de Tocqueville, *Democracy in America*, Vol. 1, ed. Phillips Bradley, trans. Henry Reeve (New York: Vintage, 1990), 99.

[25] Ibid., 101.

[26] *Moby-Dick*, 7.

[27] Ahab himself, by his own reckoning, is "the Fates' lieutenant." *Moby-Dick*, 561.

[28] Ibid., 68.

[29] Ibid., 214–15.

[30] Howard Vincent, *The Trying-out of Moby-Dick* (Boston: Houghton Mifflin, 1949), 3.

[31] "In virtually ignoring Ishmael, I am departing from a tradition in Melville criticism that has focused on Ishmael not only as the center of *Moby-Dick* but, in one way or another, as its figure of redemption." Dimock, *Empire for Liberty*, 236, n.52. Yet *by* "virtually ignoring Ishmael," Dimock makes no real attempt to suggest Melville and Ishmael are not one and the same.

[32] Dimock, *Empire for Liberty*, 111–12.

[33] Friedrich Nietzsche, *Twilight of the Idols*, in *The Portable Nietzsche*, trans. Walter Kaufmann (New York: Viking, 1968), 499.

[34] Nietzsche, *On the Genealogy of Morals*, 58–60.

[35] See Foucault, *Discipline and Punish*, especially 135–94.

[36] Dimock, *Empire for Liberty*, 115.

[37] Dimock, *Empire for Liberty*, 136.
[38] *Moby-Dick*, 201.
[39] "The Chart," like many of the chapters of *Moby-Dick* presents images that Ishmael himself is unlikely to have witnessed first hand (e.g. here, of Ahab locked in his cabin and pouring over his maps in solitude); as noted in Chapter 4 above, the Ishmael of "Loomings" cannot be the same narrator of "The Chart."
[40] Arac, "'A Romantic Book': *Moby-Dick* and Novel Agency," *boundary 2* 17.2 (Summer 1990), 47–48.
[41] *Moby-Dick*, 561; Dimock, *Empire for Liberty*, 131.
[42] *Moby-Dick*, 545.
[43] Ibid., 202, emphasis added.
[44] Arac, "'A Romantic Book'," 51, 54.
[45] *Moby-Dick*, 163–64.
[46] Dimock, *Empire for Liberty*, 120.
[47] *Moby-Dick*, 163–64.
[48] See, e.g. Nietzsche, *Thus Spoke Zarathustra*, trans. Walter Kaufmann (New York: Penguin, 1966), 99–102.
[49] *Moby-Dick*, 169.
[50] Perhaps starting with Lawrance Thompson, *Melville's Quarrel with God* (Princeton: Princeton University Press, 1952).
[51] See *Moby-Dick*, 41–48.
[52] Ibid., 164, 47.
[53] *Moby-Dick*, 140, 445; Nietzsche, *The Twilight of the Idols*, 500.
[54] *The Twilight of the Idols*, 500–01.
[55] C. L. R. James, *Mariners, Renegades, and Castaways: The Story of Herman Melville and the World We Live In* (Hanover, New Hampshire: University Press of New England, 2001), 11.
[56] Deleuze and Guattari. *A Thousand Plateaus*, trans. Brian Massumi (Minneapolis: University of Minnesota Press, 1987), 479.
[57] *Moby-Dick*, 62–63, 116.

Chapter 6

[1] Herman Melville, *Moby-Dick, or, The Whale*, eds Harrison Hayford, Hershel Parker, and G. Thomas Tanselle (Evanston and Chicago: Northwestern University Press and the Newberry Library, 1988), 3, 456.
[2] George Stewart, "The Two *Moby-Dicks*," *American Literature* 25.4 (January 1954), 418–19.
[3] Stewart relied mostly on what he called "internal evidence" (418)—i.e. close reading of the text—to develop his theory, and Harrison Hayford has since debunked it by showing that the early chapters were revised many times in the course of Melville's writing. See Hayford, "Unnecessary Duplicates: A Key to the Writing of *Moby-Dick*," in *New Perspectives on Melville*, ed. Faith Pullin (Kent: Kent State University Press, 1978), 128–161.
[4] Edward Said, "Introduction to *Moby-Dick*," in *Reflections on Exile and Other Essays* (Cambridge: Harvard University Press, 2000), 369.

[5] Melville, "Hawthorne and His Mosses," in *The Piazza Tales and Other Prose Pieces, 1839–1860*, eds Harrison Hayford, Alma A. MacDougall, G. Thomas Tanselle et al. (Evanston and Chicago: Northwestern University Press and the Newberry Library, 1987), 244.

[6] *Moby-Dick*, 198.

[7] Ibid., 203.

[8] Among the examples that leap to mind, Queequeg's tattooing and the transcription of his tattoos onto his "coffin" or canoe, the narrator's comment (in "A Bower in the Arsacides") that, in taking down the measurements of a whale on his arm, the only parchment available, he was "crowded for space, and wished the other parts of my body to remain a blank page for a poem I was composing—at least, what untattooed parts might remain" (451), and most notably perhaps, the great blankness of the whale itself, on whose blank page is inscribed, literally, a history of its battles (e.g. in "The Blanket"), and figuratively, the multitude of meanings, some of which are mentioned in "The Whiteness of the Whale."

[9] Ibid., xv.

[10] Ibid., 145, 264.

[11] Melville, *Correspondence*, eds Harrison Hayford, Alma A. MacDougall, G. Thomas Tanselle et al. (Evanston and Chicago: Northwestern University Press and the Newberry Library, 1987), 138. In the same letter to Lemuel Shaw, Melville says of *Redburn* and *White-Jacket* "my only desire for their 'success' (as it is called) springs from my pocket, & not from my heart. So far as I am individually concerned, & independent of my pocket, it is my earnest desire to write those sort of books which are said to 'fail'" (139).

[12] See Derek Gregory, *Geographical Imaginations* (Oxford: Blackwell, 1994), 73.

[13] Conrad, "Geography and Some Explorers," in *Last Essays* (London: J. M. Dent, 1926), 19–20.

[14] Conrad, *A Personal Record* (London: Heinemann, 1921), 203–204.

[15] "[M]ap-gazing, to which I became addicted so early, brings the problems of the great spaces of the earth into stimulating and direct contact with sane curiosity and gives an honest precision to one's imaginative faculty. And the honest maps of the nineteenth century nourished in me a passionate interest in the truth of geographical facts and a desire for precise knowledge which was extended later into other subjects." Conrad, "Geography and Some Explorers," 19.

[16] Conrad, *Heart of Darkness* (New York: Bantam, 1969), 11.

[17] This may be said to mark the turn, in Conrad's work, from realism to modernism, insofar as the narrative framing or "strategy of containment" (in Fredric Jameson's phrase) serves to divest the narrative of its overtly political or historical content in favor of a more existential or personal voyage of Marlowe himself. It is not that *Heart of Darkness* is exemplary of that high modernist aesthetic as much as it points the way. "After the peculiar heterogeneity of the moment of Conrad, a high modernism is set in place [. . .] The perfected poetic apparatus of high modernism represses History just as successfully as the perfected narrative apparatus of high realism did the random heterogeneity of the as yet uncentered subject. At this point, however, the political, no longer visible in the high modernist texts, any more than in the everyday world of appearance of bourgeois life, and relentlessly driven underground by accumulated reification,

has at last become a genuine Unconscious." Jameson, *The Political Unconscious* (Ithaca: Cornell University Press, 1981), 280.

18 "Once knowledge can be analyzed in terms of region, domain, implantation, displacement, transposition, one is able to capture the process by which knowledge functions as a form of power and disseminates the effects of power. There is an administration of knowledge, a politics of knowledge, relations of power which pass via knowledge and which, if one tries to transcribe them, lead one to consider forms of domination designated by such notions as field, region and territory." Michel Foucault, "Questions on Geography," in *Power/Knowledge*, ed. Colin Gordon (New York: Pantheon, 1980), 69.

19 Rebasa, *Inventing America* (Norman: University of Oklahoma Press, 1993), 183.

20 On this aspect Melville's South Seas writings, see Geoffrey Sanborn, *The Sign of the Cannibal* (Durham: Duke University Press, 1998).

21 Quoted in Carl Friedrich, *The Age of the Baroque* (New York: Harper and Row, 1952), 6.

22 On the relationship between utopia and the formation of the nation-state, see Phillip E. Wegner, *Imaginary Communities: Utopia, the Nation, and the Spatial Histories of Modernity* (Berkeley: University of California Press, 2002).

23 "The conquest of the earth, which mostly means the taking it away from those who have a different complexion or slightly flatter noses than ourselves, is not a pretty thing when you look into it too much. What redeems it is the idea only. An idea at the back of it; not a sentimental pretense but an idea; and an unselfish belief in the idea—something you can set up, and bow down before, and offer sacrifice to . . ." Conrad, *Heart of Darkness*, 9.

24 "As we look back at the cultural archive, we begin to reread it not univocally but *contrapuntally*, with a simultaneous awareness both of the metropolitan history that is narrated and of those other histories against which (and together with which) the dominating discourse acts. In the counterpoint of Western classical music, various themes play off one another, with only a provisional privilege being given to any particular one; yet in the resulting polyphony there is concert and order, an organized interplay that derives from the themes, not from a rigorous melodic or formal principle outside the work." Said, *Culture and Imperialism* (New York: Knopf, 1993), 51.

25 A telling example may be found in Melville's scathing line stating that the "small remnant of the natives [of Honolulu] had been civilized into draught horses, and evangelized into beasts of burden." *Typee*, 196. Melville suggests here and elsewhere that the savagery or barbarism of the populations on the islands is the direct result of a civilizing mission, rather than the prerequisite for it.

26 Noticing the strange tattoos on John Paul Jones's arm, Israel Potter muses: "broaches and finger-rings, not less than nose-rings and tattooing, are tokens of the primeval savageness which ever slumbers in human kind, civilized or uncivilized." Melville, *Israel Potter*, 63. *Moby-Dick*'s Queequeg is the most obvious example; a South Seas "cannibal," he can nevertheless be compared to no lesser a figure than the most heroic of all Americans: "Queequeg was George Washington cannibalistically developed." *Moby-Dick*, 50.

27 Review of *Typee*, *The New York Evangelist*, 9 April 1846. See Jay Leyda, *A Melville Log*, vol 1, 211.

[28] Melville, *Moby-Dick*, 55.

[29] Melville, *Redburn: His First Voyage*, eds Harrison Hayford, Hershel Parker, and G. Thomas Tanselle (Evanston and Chicago: Northwestern University Press and the Newberry Library, 1969), 141–50, 152.

[30] See Pratt, *Imperial Eyes: Travel Writing and Transculturation* (New York: Routledge, 1992).

[31] Melville, "The South Seas," in *The Piazza Tales and Other Prose Pieces, 1839–1860*, eds Harrison Hayford, Alma A. MacDougall, G. Thomas Tanselle et al. (Evanston and Chicago: Northwestern University Press and the Newberry Library, 1987), 410–11.

[32] Ibid., 413.

[33] See Leyda, *A Melville Log*, 196, emphasis added. There is a question here of what "real value" could mean to the editorial council at Harper and Brothers. Apparently the "first copy reader" liked *Typee* very much, asserting that "this work if not as good as Robinson Crusoe seems to me not far behind it." But "real value" was also determined by the pressures of the market; Harper leaped at the chance to publish *Omoo*, "a new work from the graphic pen of Mr. Melville, of Typee celebrity." *A Melville Log*, 230, 235.

[34] See Leyda, *A Melville Log*, 220.

[35] Ibid., 211.

[36] Twentieth-century Melville criticism and scholarship has had a hard time dealing with this matter. Early biographers, such as Raymond Weaver, took *Typee*, for instance, more or less at its word, and read the personal narratives as straightforward, autobiographical accounts. Herschel Parker has explicitly criticized this practice, but even his monumental biography, *Herman Melville*, quotes the personal narratives at length as evidence of Melville's lived experiences. But an essentially celebratory stance towards Melville required that he be seen as a great *literary* artist, who does not confine his art to facts. Hence the dilemma for such idolatrous critics: Melville's personal narratives must be both true accounts (to ensure their authority) *and* fictional or romantic (to ensure that they be valued as imaginative works of literature).

[37] *The Blithedale Romance* (New York: Dell, 1960), 21–22.

[38] In *The Marble Faun*, Hawthorne discovers that desirable "Faery Land" in Italy and asserts that the romance may escape the "insisted upon" actualities of life. In the preface to *The House of the Seven Gables*, Hawthorne distinguishes between the Romance and the Novel: the romancer "wishes to claim a certain latitude" unavailable to the novelist, who "is presumed to aim at fidelity, not merely to the possible, but to the probable and ordinary course of man's experience." The romancer is not so limited; if he be wise, he will not abuse this right, but he "can hardly be said, however, to commit a literary crime, even if he disregard this caution." Hawthorne, *The House of the Seven Gables* (Oxford: Oxford University Press, 1991), 1.

[39] Herbert, *Marquesan Encounters* (Cambridge: Harvard University Press, 1980), 183–184.

[40] This aspect of personal narrative is precisely the point in such overtly political works as Frederick Douglass's *Narrative*. Copies of Melville's own *White-Jacket* were distributed to members of Congress as they debated whether to abolish the U.S. Navy's practice of flogging.

[41] Melville, "Hawthorne and His Mosses," in *The Piazza Tales and Other Prose Pieces, 1839–1860*, eds Harrison Hayford, Alma A. MacDougall, G. Thomas Tanselle et al. (Evanston and Chicago: Northwestern University Press and the Newberry Library, 1987), 246.

[42] Franco Moretti, *Modern Epic: The World System from Goethe to García Márquez*, trans. Quentin Hoare (London: Verso, 1994), 50.

[43] See Peter Turchi, *Maps of the Imagination: The Writer as Cartographer* (San Antonio: Trinity University Press, 2004).

[44] Jameson, *Postmodernism, or, the Cultural Logic of Late Capitalism* (Durham: Duke University Press, 1991), 51. Subsequent quotations in this paragraph may be found on p. 52.

[45] Indeed, as I argued in Chapter 4, "Call me Ishmael" is not the proper point of departure for *Moby-Dick* in the first place, as the unnumbered chapters preceding it had already established a multinational frame and a postnational image.

[46] On the unintended political consequences of Mercator's projection, including its "flattering" aggrandizement of the British Empire (in drawing the prime meridian through Greenwich and making Canada look much larger than it is) and its uses by Cold War anticommunists (who liked making Russia and China look like even "bigger" threats), see Mark Monmonier, *How to Lie with Maps* (Chicago: University of Chicago Press, 1991), 94–99.

[47] Melville, *Correspondence*, 213.

[48] See "Benito Cereno," in *The Piazza Tales*, 116–17. For their notion of "the multitude" as a force against "empire," see Michael Hardt and Antonio Negri, *Empire* (Cambridge: Harvard University Press, 2000), especially 393–413.

Chapter 7

[1] Melville's enthusiasm for *Mardi* is evident on its own pages, but as more and more unflattering reviews appeared, Melville's mood seems to have shifted from amusement to defensiveness and, eventually, to a sort of acquiescence. Melville writes to Evert Duyckinck that he "thanked God it [*Mardi*] was off my hands," and in a letter to Richard Bentley, Melville jabs at the British reviewers: "the peculiar thoughts & fancies of a Yankee upon politics & other matters could hardly be presumed to delight that class of gentlemen who conduct your leading journals." But fifteen years later Melville could write of a book—Jean Paul Richter's *Titan*—that he was reviewing: "The worst thing I can say about it is that it is a little better than 'Mardi.'" See Jay Leyda, *A Melville Log: A Documentary Life of Herman Melville, 1819–1891*, Vol. 1 (New York: Harcourt, 1951), 296, 206, 668.

[2] See Jonathan Arac, *The Emergence of American Literary Narrative, 1820–1860* (Cambridge: Harvard University Press, 2005), 76. Arac contrasts these nineteenth-century personal narratives both to the more psychological novel or autobiography and to earlier forms of personal narrative, such as those by Mary Rowlandson or Jonathan Edwards, whose main purpose was to show the spiritual and religious transformations of the individual narrator.

[3] Obviously this does not apply to the slave narratives of the period, whose general movement was one of ascent, from the hell of slavery to the land of freedom (upward and northward). From a purely dramatic standpoint, however, one can still see elements of this descent and return, as even slave narratives often present a nadir lower than the initial position of the narrator. For example, in Frederick Douglass's 1845 *Narrative*, Douglass starts out as a slave on the Great House Farm before moving to Baltimore—where "A city slave is practically a freeman compared with a slave on the plantation"—only to be cast down into the more abject condition of working for Mr. Covey, a noted slave "breaker," under whose baleful influence Douglass became "a man transformed into a brute." Only after his fistfight with Covey ("the turning-point in my career as a slave") was Douglass "resurrected," and from there made his way to physical freedom, having emancipated himself spiritually. Chapter X, which recounts his time with Covey, takes up more than a third of the whole *Narrative*. See Douglass, *Narrative* (New York: Dover, 1995), 21, 34, 38, 43. For a comparison of the slave narrative form and Melville's personal narrative, see John Carlos Rowe, "Melville's *Typee*: U.S. Imperialism at Home and Abroad," in *National Identities and Post-Americanist Narratives*, ed. Donald E. Pease (Durham: Duke University Press, 1994), 255–78.

[4] Herman Melville, *Redburn: His First Voyage*, eds Harrison Hayford, Hershel Parker, and G. Thomas Tanselle (Evanston and Chicago: Northwestern University Press and the Newberry Library, 1969), 50–52.

[5] Melville, *Correspondence*, eds Harrison Hayford, Alma A. MacDougall, G. Thomas Tanselle et al. (Evanston and Chicago: Northwestern University Press and the Newberry Library, 1987), 114–15.

[6] Ibid., 106.

[7] Ibid., 106.

[8] Franco Moretti, however, makes no mention of Poe's cosmological "prose poem" or of *Mardi* in his discussion of "world texts" in *Modern Epic*.

[9] Melville, *Mardi: A Voyage Thither*, eds Harrison Hayford, Hershel Parker, and G. Thomas Tanselle (Evanston and Chicago: Northwestern University Press and the Newberry Library, 1970), 13.

[10] According to the fiction of *Mardi*, all of the story is set in the South Seas, even when (as readers are quick to recognize) the place described is clearly not located there. For example, Vivenza looks an awful lot like the United States, Franko like France, etc.

[11] Arac, *Emergence*, 140.

[12] In a letter to Hawthorne, Melville complains about the source of his renown. "All Fame is patronage. Let me be infamous: there is no patronage in *that*. What "reputation" H.M. has is horrible. Think of it! To go down to posterity is bad enough, any way; but to go down as a "man who lived among the cannibals"! When I speak of posterity, in reference to myself, I only mean the babies who will probably be born in the moment immediately ensuing upon my giving up the ghost. I shall go down to some of them, in all likelihood. "Typee" will be given to them, perhaps, with their gingerbread." Melville, *Correspondence*, 193.

[13] Fredric Jameson, *Postmodernism, or, The Cultural Logic of Late Capitalism* (Durham: Duke University Press, 1991), 52.

[14] Melville, *Correspondence*, 138–39.

[15] Ibid., 132.

[16] Melville, *White-Jacket, or, The World in a Man-of-War*, eds Harrison Hayford, Hershel Parker, and G. Thomas Tanselle (Evanston and Chicago: Northwestern University Press and the Newberry Library, 1970), 74.

[17] Arac, *Commissioned Spirits* (New York: Columbia University Press, 1989), 5, 7.

[18] This, of course, is very much Foucault's point in his critique of power (and of various theories of power) in *Discipline and Punish*. "We must cease once and for all to describe the effects of power in negative terms: it "excludes," it "represses," it "censors," it "abstracts," it "masks," it "conceals." In fact, power produces; it produces reality; it produces domains of objects and rituals of truth." Foucault, *Discipline and Punish*, trans. Alan Sheridan (New York: Vintage, 1977), 194.

[19] *Redburn*, 5.

[20] On this term, see Saskia Sassen, *The Global City: New York, London, Tokyo* (Princeton: Princeton University Press, 1991).

[21] Karl Marx and Friedrich Engels, *The Communist Manifesto*, trans. anon. (New York: Signet, 1998), 55.

[22] A humorous shipboard episode in *Redburn* recounts Redburn's struggle to make sense of Adam Smith's *The Wealth of Nations*, believing that its wisdom will reveal the solution to his family's poverty. Redburn begins reading the book, "where I fancied lay something like the philosopher's stone, a secret talisman, which would transmute even pitch and tar to silver and gold. [. . .] So I read on and on, about "*wages and profits of labor*," without getting any profits myself for my pains in perusing it." He eventually stops reading, wraps his jacket around it, and uses the tome for a pillow, "for which purpose it answered well; only I sometimes waked up feeling dull and stupid; but of course the book could not have been the cause of that." *Redburn*, 86–87.

[23] See Arac, *Emergence*, 71.

[24] See Melville, *Pierre, or, The Ambiguities*, eds Harrison Hayford, Hershel Parker, and G. Thomas Tanselle (Evanston and Chicago: Northwestern University Press and the Newberry Library, 1995), 208.

[25] Poe, "The Man of the Crowd," in *The Fall of the House of Usher and Other Writings*, ed. David Galloway (New York: Penguin, 1986), 140.

[26] Wyn Kelley, *Melville's City* (Cambridge: Cambridge University Press, 1996), 68–69.

[27] Raymond Williams, *The Country and the City* (Oxford: Oxford University Press, 1973), 165.

[28] *Redburn*, 152. The other quotations in this paragraph are from pages 152 and 157, in the chapter titled "With his Prosy Old Guide-Book, He Takes a Prosy Stroll through the Town."

[29] *Legend*, from the Latin, meaning something "to be read," and the *legend* that appears on a map is understood as the key to understanding it. The chapters that follow the guide-book discussions could be seen as Redburn's attempt to rewrite the guide-book, to update it. But, as Melville's "prosy stroll" chapter peremptorily warns, guide-books are "the least reliable" of all books.

[30] *Redburn*, 159.

[31] Jameson, *Postmodernism*, 51.

[32] An alternative view is found in Baudelaire's notion of the *flâneur,* the leisurely stroller who moves among the urban throng, observing the rhythmic chaos of the quotidian. As Walter Benjamin notes, however, Baudelaire's equation of "the man of the crowd" with the *flâneur* is misleading, since the former is a "manic" pedestrian "jostled by the crowd" and the latter is the "man of leisure" who wandering among the crowd without being part of it. "Let the many attend to their daily affairs, the man of leisure can indulge in the perambulations of the *flâneur* only if as such he is already out of place." Benjamin, "On Some Motifs in Baudelaire," in *Illuminations: Essays and Reflections,* trans. Harry Zohn (New York: Schocken, 1985), 172.

[33] Michel de Certeau, *The Practice of Everyday Life,* trans. Steven Rendall (Berkeley: University of California Press, 1984), 91.

[34] Quoted in Benjamin, "On Some Motifs in Baudelaire," 175.

[35] De Certeau, The Practice of Everyday Life, 101.

[36] Georg Lukács, "Narrate or Describe," in *Writer and Critic,* trans. Arthur D. Kahn (New York: Grosset, 1970), 120, 130.

[37] On the use of actual maps and guides in establishing realism in literature, see Eric Bulson, *Novels, Maps, Modernity: The Spatial Imagination, 1850–2000* (London: Routledge, 2007).

[38] *Redburn,* 227.

[39] Jameson also makes this point regarding "cognitive mapping," or "self-consciousness about the social totality": "My thesis, however, is not merely that we ought to strive for it, but that we do so all the time anyway without [necessarily] being aware of the process." *The Geopolitical Aesthetic* (Bloomington: Indiana University Press, 1992), 2.

Chapter 8

[1] See Saskia Sassen, *The Global City: New York, London, Tokyo* (Princeton: Princeton University Press, 1991).

[2] Herman Melville, *Moby-Dick, or, The Whale,* eds Harrison Hayford, Hershel Parker, and G. Thomas Tanselle (Evanston and Chicago: Northwestern University Press and the Newberry Library, 1988), 3.

[3] Raymond Williams, *The Country and the City* (Oxford: Oxford University Press, 1973), 279.

[4] Melville, *Correspondence,* eds Harrison Hayford, Alma A. MacDougall, G. Thomas Tanselle et al. (Evanston and Chicago: Northwestern University Press and the Newberry Library, 1987), 218–19.

[5] In "The Symphony," Melville extends the metaphor, identifying the air as the feminine realm and the sea as masculine. The "transparently soft and pure" air is filled with birds that "were the gentle thoughts of the feminine air," while "mighty leviathans, sword-fish, and sharks [. . .] were the strong, troubled, murderous thinkings of the masculine sea." At the surface, however, "the two seemed one; it was only the sex, as it were, that distinguished them." *Moby-Dick,* 542.

[6] Arac, *The Emergence of American Literary Narrative, 1820–1860* (Cambridge: Harvard University Press, 2005), 73.

[7] The distinction between the *house* and the *railroad* is elaborated in Jonathan Arac, *Commissioned Spirits: The Shaping of Social Motion in Dickens, Carlyle, Melville, and Hawthorne*. New York: Columbia University Press, 1989), 94–113. While this particular conflict does not appear directly in *Pierre*—there is no railroad in the novel—it still figures in the rural/urban opposition, as is most dramatically visible in Hawthorne's *The House of the Seven Gables*. Like *Pierre*, Hawthorne's novel associates tradition with the home, a domestic space *haunted* by tradition; and though *Pierre* lacks a train-ride, Clifford's and Hepzibah's wild voyage leading nowhere resembles Pierre's and Isabel's carriage-ride into New York.

[8] Heidegger, *Being and Time*, trans. John Macquarrie and Edward Robinson (New York: Harper and Row, 1962), 233. Of course, the canonical statement of the eerie feeling and anxiety associated with "the uncanny" (*das Unheimliche*) may be found in Sigmund Freud, "The Uncanny," trans. Alix Strachey, in *Studies in Parapyschology*, ed. Philip Rieff (New York: Collier, 1963).

[9] Melville, *Pierre, or, The Ambiguities*, eds Harrison Hayford, Hershel Parker, and G. Thomas Tanselle (Evanston and Chicago: Northwestern University Press and the Newberry Library, 1995), 3.

[10] Ibid., 5, 13.

[11] Ibid., 13.

[12] Ibid., 12–13.

[13] Ibid., 19–20.

[14] Ibid., 18.

[15] Spanos, *Herman Melville and the American Calling: The Fiction After "Moby-Dick," 1851–1857* (Albany: State University of New York Press, 2008), 23.

[16] As Raymond Williams summarizes, "The concept of hegemony sees the relations of domination and subordination in the, in their forms of practical consciousness, as in effect a saturation of the whole process of living—not only in political and economic activity, nor only of manifest social activity, but of the whole substance of lived identities and relationships, to such a depth that the pressures and limits of what can ultimately be seen as a specific economic, political, and cultural system seem to most of us the pressures and limits of simple experience and common sense. Hegemony [. . .] is a lived system of meanings and values—constitutive and constituting—which as they are experienced as practices appear as reciprocally confirming. It thus constitutes a sense of reality for most people in the society, a sense of absolute because experienced reality beyond which it is very difficult for most members of the society to move, in most areas of their lives." Williams, *Marxism and Literature* (Oxford: Oxford University Press, 1977), 110.

[17] *Pierre*, 30.

[18] Ibid., 20.

[19] Michel Foucault, *Discipline and Punish*, trans. Alan Sheridan (New York: Vintage, 1977), 169.

[20] Spanos, Herman Melville and the American Calling, 52.

[21] *Pierre*, 171.

[22] Poe, "The Man of the Crowd," in *The Fall of the House of Usher and Other Writings*, ed. David Galloway (New York: Penguin, 1986), 179.

[23] "Bartleby, the Scrivener," in *The Piazza Tales and Other Prose Pieces, 1839–1860*, ed. Harrison Hayford, Alma A. MacDougall, G. Thomas Tanselle et al. (Chicago and Evanston: The Newberry Library and Northwestern University Press, 1987), 14.

[24] Ibid., 19–20.

[25] See Leo Marx, "Melville's Parable of the Walls." *The Sewanee Review* 61 (October 1953): 602–627.

[26] James C. Wilson, "'Bartleby': The Walls of Wall Street," *Arizona Quarterly* 27 (Winter 1981): 338; Louise K. Barnet, "Bartleby as Alienated Worker," *Studies in Short Fiction* 11 (1974): 379.

[27] C. L. R. James, *Mariners, Renegades, and Castaways: The Story of Herman Melville and the World We Live In* (Hanover, New Hampshire: University Press of New England, 2001), 89.

[28] Gilles Deleuze, "Bartleby, or the Formula," in *Essays Critical and Clinical*, trans. Daniel W. Smith and Michael A. Greco (Minneapolis: University of Minnesota Press, 1997), 73.

[29] See my "Reading the Original: Alienation, Writing, and Labor in 'Bartleby, the Scrivener,'" in *Bloom's Literary Themes: Alienation*, eds Harold Bloom and Blake Hobby (New York: Chelsea House, forthcoming 2009).

[30] "Bartleby," 25.

[31] Ibid., 37.

[32] Ibid., 44.

[33] *Moby-Dick*, 164.

[34] Ibid., 44.

Conclusion

[1] Melville, *Moby-Dick, or, The Whale*, eds Harrison Hayford, Hershel Parker, and G. Thomas Tanselle (Evanston and Chicago: Northwestern University Press and the Newberry Library, 1988), 278–81.

[2] Jonathan Arac, *Commissioned Spirits: The Shaping of Social Motion in Dickens, Carlyle, Melville, and Hawthorne* (New York: Columbia University Press, 1989), 35.

[3] Gilles Deleuze and Claire Parnet, *Dialogues*, trans. Hugh Tomlinson and Barabara Habberjam (New York: Columbia University Press, 1987), 36.

[4] See Kenichi Ohmae, *The Borderless World: Power and Strategy in the Interlinked Economy* (New York: Harper Collins, 1990).

[5] Happily, a rather long list could be made. See, for just one recent example, Thomsen, Mads Rosendahl Thomsen, *Mapping World Literature: International Canonization and Transnational Literatures* (London: Continuum Books, 2008).

[6] Erich Auerbach, "Philology and *Weltliteratur*," trans M. and E. W. Said. *Centennial Review* 13 (Winter 1969): 17.

[7] C. L. R. James, *Mariners, Renegades, and Castaways: The Story of Herman Melville and the World We Live In* (Hanover, New Hampshire: University Press of New England, 2001), 20.

[8] See Samuel Eliot Morison, *The Maritime History of Massachusetts, 1783–1860* (Boston: Houghton Mifflin, 1941).

[9] Jameson continues: "The truth of that limited daily experience of London lies, rather, in India or Jamaica or Hong Kong; it is bound up with the whole colonial system of the British Empire that determines the very quality of the individual's subjective life. Yet those structural coordinates are no longer even accessible to immediate lived experience and are often not even conceptualizable for most people." Jameson, *Postmodernism, or, The Cultural Logic of Late Capitalism* (Durham: Duke University Press, 1991), 411.

[10] See Edward LiPuma and Benjamin Lee, *Financial Derivatives and the Globalization of Risk* (Durham: Duke University Press, 2004), 3.

[11] Jameson, *Postmodernism*, 54.

[12] Melville, *Correspondence*, eds Harrison Hayford, Alma A. MacDougall, G. Thomas Tanselle et al. (Evanston and Chicago: Northwestern University Press and the Newberry Library, 1987), 213.

Bibliography

Althusser, Louis (1971) Ideology and Ideological State Apparatuses. In Ben Brewster (trans.), *Lenin and Philosophy and Other Essays.* New York: Monthly Review Press, pp. 127–86.

Anderson, Benedict (1991) *Imagined Communities: Reflections on the Origin and Spread of Nationalism.* London: Verso.

Arac, Jonathan (1985) F. O. Matthiessen: Authorizing the American Renaissance. In Walter Benn Michaels and Donald E. Pease (eds), *The American Renaissance Reconsidered.* Baltimore: Johns Hopkins University Press, pp. 90–112.

— (1986) The Politics of *The Scarlet Letter.* In Sacvan Bercovitch and Myra Jehlen (eds), *Ideology and Classic American Literature.* Cambridge: Cambridge University Press, pp. 247–66.

— (1989) *Commissioned Spirits: The Shaping of Social Motion in Dickens, Carlyle, Melville, and Hawthorne.* New York: Columbia University Press.

— (1990) "A Romantic Book": *Moby-Dick* and Novel Agency. *boundary 2* 17.2, (Summer): 40–59.

— (1995) Narrative Forms. In Sacvan Bercovitch (ed.) *The Cambridge History of American Literature, Vol. 2: 1820–1860.* Cambridge: Cambridge University Press, 1995.

— (1997) *Huckleberry Finn as Idol and Target: The Functions of Criticism in Our Time.* Madison: University of Wisconsin Press, 1997.

— (2005) *The Emergence of American Literary Narrative, 1820–1860.* Cambridge: Harvard University Press, 2005.

— (2007) Global or Babel: Language and Planet in American Literature. In Wai Chee Dimock and Lawrence Buell (eds), *Shades of the Planet.* Princeton: Princeton University Press, pp. 19–38.

Armstrong, Philip (2004) "Leviathan is a Skein of Networks": Translations of Nature and Culture in *Moby-Dick. ELH* 71: 1039–63.

Auerbach, Erich (1969) Philology and *Weltliteratur,* trans M. and E. W. Said. *Centennial Review* 13 (Winter): 1–17.

Barnett, Louise K. (1974) Bartleby as Alienated Worker. *Studies in Short Fiction* 11: 379–385.

Baudelaire, Charles (1972) The Painter of Modern Life. In P. E. Charvet (trans.) *Selected Writings on Art and Literature.* New York: Vintage, pp. 395–422.

Bakhtin, Mikhail (1981) *The Dialogic Imagination,* trans Caryl Emerson and Michael Holquist. Austin: University of Texas Press.

Benjamin, Walter (1977) *The Origin of German Tragic Drama,* trans. John Osborne. London: Verso.

— (1985) On Some Motifs in Baudelaire. In Harry Zohn (trans.) *Illuminations: Essays and Reflections.* New York: Schocken, pp. 155–200.

Bercovitch, Sacvan (1978) *The American Jeremiad*. Madison: University of Wisconsin Press.

— (1993) *The Rites of Assent: Transformations in the Symbolic Construction of America*. London: Routledge.

— and Jehlen, Myra, eds (1986) *Ideology and Classic American Literature*. Cambridge: Cambridge University Press.

Berlin, Isaiah (1998) *The Crooked Timber of Humanity*. Princeton: Princeton University Press.

Beverley, John (1993) *Against Literature*. Minneapolis: University of Minnesota Press.

Bloom, Harold (2007) Introduction. In Robert T. Tally Jr. (vol. ed.) *Bloom's Classic Critical Views: Edgar Allan Poe*. New York: Chelsea House, pp. xi–xii.

Borges, Jorge Luis (1975) Of Exactitude in Science. In Norman Thomas de Giovanni (trans.) *A Universal History of Infamy*. London: Penguin.

Bulson, Eric (2007) *Novels, Maps, Modernity: The Spatial Imagination, 1850–2000*. London: Routledge.

Casanova, Pascale (2004). *The World Republic of Letters*, trans. M. B. DeBevoise. Cambridge: Harvard University Press.

Casarino, Cesare (2002). *Modernity at Sea: Marx, Melville and Conrad in Crisis*. Minneapolis: University of Minnesota Press.

Certeau, Michel de (1984) *The Practice of Everyday Life*, trans. Steven Rendall. Berkeley: University of California Press.

Conrad, Joseph (1921) *A Personal Record*. London: Heinemann.

— (1926) Geography and Some Explorers. In *Last Essays*. London: J. M. Dent, pp. 1–31.

— (1969) *Heart of Darkness*. New York: Bantam.

Cooper, James Fenimore (1986) *The Last of the Mohicans*. New York: Penguin.

Deleuze, Gilles (1985) Nomad Thought, trans. David B. Allison. In David B. Allison (ed.) *The New Nietzsche*. Cambridge: MIT Press.

— (1988) *Spinoza: Practical Philosophy*, trans. Robert Hurley. San Francisco: City Lights.

— (1990) *Expressionism in Philosophy: Spinoza*, trans. Martin Joughin. New York: Zone Books.

— (1993) *The Fold: Leibniz and the Baroque*, trans. Tom Conley. Minneapolis: University of Minnesota Press.

— (1994) *Difference and Repetition*, trans. Paul Patton. New York: Columbia University Press.

— (1997) Bartleby, or, the Formula. In Daniel W. Smith and Michael A. Greco (trans) *Essays Critical and Clinical*. Minneapolis: University of Minnesota Press, pp. 68–90.

Deleuze, Gilles and Guattari, Félix (1983) *Anti-Oedipus*, trans Robert Hurley, Mark Seem, and Helen R. Lane. Minneapolis: University of Minnesota Press.

— (1987) *A Thousand Plateaus*, trans. Brian Massumi. Minneapolis: University of Minnesota Press.

— (1994) *What is Philosophy?*, trans Hugh Tomlinson and Grant Burchell. New York: Columbia University Press.

Deleuze, Gilles, and Claire Parnet (1987) *Dialogues*, trans Hugh Tomlinson and Barabara Habberjam. New York: Columbia University Press.

Dimock, Wai Chee (1989) *Empire for Liberty: Melville and the Poetics of Individualism*. Princeton: Princeton University Press.

— (2006) *Through Other Continents: American Literature Across Deep Time*. Princeton: Princeton University Press.

Dimock, Wai Chee, and Lawrence Buell (eds) (2008) *Shades of the Planet: American Literature as World Literature*. Princeton: Princeton University Press.

Douglass, Frederick. (1995) *Narrative of the Life of Frederick Douglass, Written by Himself*. New York: Dover.

Duyckinck, Evert (1851) Melville's Moby-Dick; or, the Whale. *The Literary World* (November 22): 403–04.

Eagleton, Terry (1996) *Literary Theory* (2nd edn). Minneapolis: University of Minnesota Press.

Emerson, Ralph Waldo (1982) History. In *Selected Essays*, Larzer Ziff (ed.). New York: Penguin, pp. 149–173.

Focillon, Henri (1992) *The Life of Forms in Art*, trans Charles B. Hogan and George Kubler. New York: Zone Books.

Foucault, Michel (1973) *The Order of Things: An Archaeology of the Human Sciences*, trans anonymous collective. New York: Vintage.

— (1975) *The Birth of the Clinic: An Archaeology of Medical Perception*, trans. A. M. Sheridan Smith. New York: Vintage.

— (1977a) *Discipline and Punish: The Birth of the Prison*, trans. Alan Sheridan. New York: Vintage.

— (1977b) Nietzsche, Genealogy, History. In Donald F. Bouchard and Sherry Simon (trans), *Language, Counter-Memory, Practice*. Ithaca: Cornell University Press.

— (1980) Questions on Geography. In Colin Gordon (ed.) *Power/Knowledge*. New York: Pantheon, pp. 63–77.

Fredericks, Nancy (1995) *Melville's Art of Democracy*. Athens: University of Georgia Press.

Freud, Sigmund (1963) The Uncanny, trans. Alix Strachey. In Philip Rieff (ed.), *Studies in Parapsychology*. New York: Collier, pp. 19–60.

Friedrich, Carl J. (1952) *The Age of the Baroque, 1610–1660*. New York: Harper and Row.

Goethe, Johann Wolfgang von (1994) On World Literature. In John Geary (ed.), Ellen von Nardhoff and Ernest H. von Nardhoff (trans), *Essays on Art and Literature*. Princeton: Princeton University Press, 1994.

Goldner, Loren (2007) *Herman Melville: Between Charlemagne and the Antemosaic Cosmic Man*. New York: Queequeg Publications.

Gramsci, Antonio (1971) *Selections from the Prison Notebooks*, eds and trans Quintin Hoare and Geoffrey Nowell Smith. New York: International Publishers.

Gregory, Derek (1994). *Geographical Imaginations*. Oxford: Blackwell.

Habermas, Jürgen (2001) *The Postnational Constellation*, trans. Max Pensky. Cambridge: MIT Press.

Hardt, Michael, and Antonio Negri (2000) *Empire*. Cambridge: Harvard University Press.

Harvey, Bruce A. (2001) *American Geographics: U.S. National Narratives and the Representation of the Non-European World, 1830–1865*. Palo Alto: Stanford University Press.

Hassan, Ihab (2005) Maps and Stories: A Brief Meditation. *Georgia Review* 59.4 (Winter): 751–63.

Hawthorne, Nathaniel (1987a) Young Goodman Brown. In Brian Harding (ed.), *Young Goodman Brown and Other Tales.* Oxford: Oxford University Press, pp. 111–24.

— (1987b) Ethan Brand. In Brian Harding (ed.), *Young Goodman Brown and Other Tales.* Oxford: Oxford University Press, pp. 316–32.

— (1987c) *Letters, 1853–1856.* Columbus: The Ohio State University Press.

— (1990) *The Scarlet Letter.* Oxford: Oxford University Press.

— (1991a) *The Blithedale Romance.* Oxford: Oxford University Press.

— (1991b) *The House of the Seven Gables.* Oxford: Oxford University Press.

Hayford, Harrison (1978) Unnecessary Duplicates: A Key to the Writing of *Moby-Dick.* In Faith Pullin (ed.), *New Perspectives on Melville.* Kent: Kent State University Press, pp. 128–61.

Heidegger, Martin (1962) *Being and Time,* trans John Macquarrie and Edward Robinson. New York: Harper and Row.

Heimert, Alan (1963) *Moby-Dick* and American Political Symbolism. *American Quarterly,* 15.4 (Winter): 498–534.

Herbert, T. Walter (1980) *Marquesan Encounters.* Cambridge: Harvard University Press.

Huang, Yunte (2008) *Transpacific Imaginations: History, Literature, Counterpoetics.* Cambridge: Harvard University Press.

Irving, Washington (1961) The Legend of Sleepy Hollow. In *The Sketch Book.* New York: Penguin.

James, C. L. R. (2001) *Mariners, Renegades, and Castaways: The Story of Herman Melville and the World We Live In.* Hanover, New Hampshire: University Press of New England.

Jameson, Fredric (1971) *Marxism and Form: Twentieth-Century Dialectical Theories of Literature.* Princeton: Princeton University Press.

— (1981) *The Political Unconscious: Narrative as a Socially Symbolic Act.* Ithaca: Cornell University Press.

— (1991) *Postmodernism, or, The Cultural Logic of Late Capitalism.* Durham: Duke University Press.

— (1992) *The Geopolitical Aesthetic: Cinema and Space in the World System.* Bloomington: Indiana University Press.

— (2005) *Archaeologies of the Future: The Desire Called Science Fiction and Other Science Fictions.* London: Verso.

Kant, Immanuel (1933) *Critique of Pure Reason,* trans. Norman Kemp Smith. London: Macmillan.

Kelley, Wyn (1996) *Melville's City: Literary and Urban Form in Nineteenth-Century New York.* Cambridge: Cambridge University Press.

— (ed.) (2002) *A Companion to Herman Melville.* Oxford: Blackwell Publishing.

Lawrence, D. H. (1964) *Studies in Classic American Literature.* New York: Viking.

Lefebvre, Henri (1991) *The Production of Space,* trans. Donald Nicholson-Smith. Oxford: Blackwell.

Lévi-Strauss, Claude (1966) *The Savage Mind,* trans John and Doreen Weightman. Chicago: University of Chicago Press.

Levin, Harry (1989) *The Power of Blackness.* Athens, Ohio: Ohio University Press.

Lewis, R. W. B. (1955) *The American Adam: Innocence, Tragedy, and Tradition in the Nineteenth Century.* Chicago: University of Chicago Press.

Leyda, Jay (1951) *A Melville Log: A Documentary Life of Herman Melville, 1819–1891.* 2 vols. New York: Harcourt.

LiPuma, Edward, and Lee, Benjamin (2004) *Financial Derivatives and the Globalization of Risk.* Durham: Duke University Press.

Lukács, Georg (1970) Narrate or Describe. In Arthur D. Kahn (trans.), *Writer and Critic.* New York: Grosset, pp. 110–48.

— (1971) *The Theory of the Novel,* trans. Anna Bostock. Cambridge: MIT Press.

Lynch, Kevin (1960) *The Image of the City.* Cambridge: MIT Press.

Lyons, Paul (2002) Global Melville. In Wyn Kelley (ed.), *A Companion to Herman Melville.* Oxford: Blackwell Publishing, pp. 52–67.

— (2006) *American Pacificism: Oceania in the U.S. Imagination.* New York: Routledge.

Marin, Louis (1990) *Utopics: The Semiological Place of Textual Spaces,* trans. Robert A. Vollrath. Atlantic Highlands, N.J.: Humanities Press.

Marx, Karl (1963) *The Eighteenth Brumaire of Louis Bonaparte,* trans. anonymous. New York: International Publishers.

— (1986) Bastiat and Carey. In *Collected Works of Karl Marx and Friedrich Engels,* vol. 28, trans. Ernst Wangermann. Moscow: International Publishers, pp. 5–16.

Marx, Karl, and Friedrich Engels (1998) *The Communist Manifesto,* trans. anonymous. New York: Signet.

Marx, Leo (1953) Melville's Parable of the Walls. *The Sewanee Review* 61 (October): 602–627.

— (1964) *The Machine in the Garden: Technology and the Pastoral Ideal in America.* Oxford: Oxford University Press.

— (2005) On Recovering the "Ur" Theory of American Studies. *American Literary History* 17.1 (Spring): 118–134.

Matthiessen, F. O. (1941) *American Renaissance: Art and Expression in the Age of Emerson and Whitman.* Oxford: Oxford University Press.

Melville, Herman (1968a) *Omoo: A Narrative of Adventures in the South Seas,* eds Harrison Hayford, Hershel Parker, and G. Thomas Tanselle. Evanston and Chicago: Northwestern University Press and the Newberry Library.

— (1968b) *Typee: A Peep at Polynesian Life,* eds Harrison Hayford, Hershel Parker, and G. Thomas Tanselle. Evanston and Chicago: Northwestern University Press and the Newberry Library.

— (1969) *Redburn: His First Voyage,* eds Harrison Hayford, Hershel Parker, and G. Thomas Tanselle. Evanston and Chicago: Northwestern University Press and the Newberry Library.

— (1970a) *White-Jacket, or, The World in a Man-of-War,* eds Harrison Hayford, Hershel Parker, and G. Thomas Tanselle. Evanston and Chicago: Northwestern University Press and the Newberry Library.

— (1970b) *Mardi and A Voyage Thither,* eds Harrison Hayford, Hershel Parker, and G. Thomas Tanselle. Evanston and Chicago: Northwestern University Press and the Newberry Library.

— (1982) *Israel Potter: His Fifty Years in Exile*, eds Harrison Hayford, Hershel Parker, and G. Thomas Tanselle. Evanston and Chicago: Northwestern University Press and the Newberry Library.

— (1984) *The Confidence-Man: His Masquerade*, eds Harrison Hayford, Hershel Parker, and G. Thomas Tanselle. Evanston and Chicago: Northwestern University Press and the Newberry Library.

— (1987a) *Correspondence*, eds Harrison Hayford, Alma A. MacDougall, G. Thomas Tanselle et al. Evanston and Chicago: Northwestern University Press and the Newberry Library.

— (1987b) *The Piazza Tales and Other Prose Pieces, 1839–1860*, eds Harrison Hayford, Alma A. MacDougall, G. Thomas Tanselle et al. Evanston and Chicago: Northwestern University Press and the Newberry Library.

— (1988) *Moby-Dick, or, The Whale*, eds Harrison Hayford, Hershel Parker, and G. Thomas Tanselle. Evanston and Chicago: Northwestern University Press and the Newberry Library.

— (1989) *Journals*, eds Harrison Hayford, Hershel Parker, and G. Thomas Tanselle. Evanston and Chicago: Northwestern University Press and the Newberry Library.

— (1995) *Pierre, or, The Ambiguities*, eds Harrison Hayford, Hershel Parker, and G. Thomas Tanselle. Evanston and Chicago: Northwestern University Press and the Newberry Library.

Michaels, Walter Benn, and Pease, Donald E. (eds) (1985) *The American Renaissance Reconsidered*. Baltimore: Johns Hopkins University Press.

Miller, Perry (1956) *Errand into the Wilderness*. New York: Harper and Row.

Monmonier, Mark (1991) *How to Lie with Maps*. Chicago: University of Chicago Press.

Moretti, Franco (1994) *Modern Epic: The World System from Goethe to García Márquez*, trans. Quentin Hoare. London: Verso.

— (2005) *Graphs, Maps, Trees: Abstract Models for a Literary History*. London: Verso.

— (ed.) (2006) *The Novel*, 2 vols. Princeton: Princeton University Press.

Morison, Samuel Eliot (1941) *The Maritime History of Massachusetts, 1783–1860*. Boston: Houghton Mifflin.

Negri, Antonio (2007) *The Political Descartes: Reason, Ideology, and the Bourgeois Project*, trans Matteo Mandarini and Alberto Toscano. London: Verso.

Nietzsche, Friedrich (1968) *Twilight of the Idols*. In Walter Kaufman (trans.), *The Portable Nietzsche*. New York: Viking, pp. 463–564.

— (1969) *On the Genealogy of Morals and Ecce Homo*, trans Walter Kaufmann and R. J. Hollingdale. New York: Penguin.

— (1978) *Thus Spoke Zarathustra*, trans. Walter Kaufmann. New York: Penguin.

Ohmae, Kenichi (1990) *The Borderless World: Power and Strategy in the Interlinked Economy*. New York: Harper Collins.

Olson, Charles (1947) *Call Me Ishmael*. San Francisco: City Lights.

Parker, Hershel (2005) *Herman Melville: A Biography*. 2 vols. Baltimore: Johns Hopkins University Press, 1996.

Pease, Donald E. (1985) *Moby Dick* and the Cold War. In Walter Benn Michaels and Donald E. Pease (eds), *The American Renaissance Reconsidered*. Baltimore: Johns Hopkins University Press, pp. 113–55.

Pease, Donald E. (ed.) (1994a) *National Identities and Post-Americanist Narratives.* Durham: Duke University Press.

— (1994b) National Identities, Postmodern Artifacts, and Postnational Narratives. In Donald E. Pease (ed.), *National Identities and Post-Americanist Narratives.* Durham: Duke University Press, pp. 1–13.

Poe, Edgar Allan (1986) *The Fall of the House of Usher and Other Writings*, ed. David Galloway. New York: Penguin.

Pratt, Mary Louise (1992) *Imperial Eyes: Travel Writing and Transculturation.* New York: Routledge.

Pynchon, Thomas (1966) *The Crying of Lot 49.* New York: Harper and Row.

Rabasa, José (1993) *Inventing America: Spanish Historiography and the Formation of Eurocentrism.* Norman: University of Oklahoma Press.

Richard, Claude (1998) *American Letters*, trans. Carol Mastrangelo Bové. Philadelphia: University of Pennsylvania Press.

Rowe, John Carlos (1994). Melville's *Typee*: U.S. Imperialism at Home and Abroad. In Donald E. Pease (ed.), *National Identities and Post-Americanist Narratives.* Durham: Duke University Press, pp. 255–78.

— (ed.) (2000) *Post-Nationalist American Studies.* Berkeley: University of California Press.

Said, Edward (2000) Introduction to *Moby-Dick.* In *Reflections on Exile and Other Essays.* Cambridge: Harvard University Press, pp. 356–371.

— (1985) *Beginnings: Intention and Method.* New York: Columbia University Press.

— (1993) *Culture and Imperialism.* New York: Knopf.

Sanborn, Geoffrey (1998) *The Sign of the Cannibal: Melville and the Making of the Postcolonial Reader.* Durham: Duke University Press.

Sassen, Saskia (1991) *The Global City: New York, London, Tokyo.* Princeton: Princeton University Press.

Schleifer, Neil (1994) Melville as Lexicographer. *Melville Society Extracts* 98: 1–6.

Sealts, Merton (1966) *Melville's Reading: A Checklist of Books Owned or Borrowed.* Madison: University of Wisconsin.

Smith, Henry Nash (1950) *Virgin Land: The American West as Symbol and Myth.* Cambridge: Harvard University Press.

Soja, Edward W. (1989) *Postmodern Geographies: The Reassertion of Space in Critical Social Theory.* London: Verso.

Spanos, William V. (1995) *The Errant Art of "Moby-Dick": The Canon, the Cold War, and the Struggle for American Studies.* Durham: Duke University Press.

— (2001) Pierre's Extraordinary Emergency: Melville and "The Voice of Silence," Part I. *boundary 2* 28.2 (Summer): 105–31.

— (2008) *Herman Melville and the American Calling: The Fiction After "Moby-Dick," 1851–1857.* Albany: State University of New York Press.

Spengemann, William (1996) Introduction. In Herman Melville, *Pierre.* New York: Penguin, pp. vii–xx.

Stewart, George (1954) The Two *Moby-Dicks. American Literature* 25.4 (January): 417–448.

Tally, Robert T., Jr. (2006) "Believing in America": The Politics of American Studies in a Postnational Era. *The Americanist* XXIII: 69–81.

— (vol. ed.) (2007) *Bloom's Classic Critical Views: Edgar Allan Poe.* New York: Chelsea House.

— (2009) Reading the Original: Alienation, Writing, and Labor in "Bartleby, the Scrivener." In Harold Bloom and Blake Hobby (eds), *Bloom's Literary Themes: Alienation.* New York: Chelsea House.

Thompson, Lawrance (1952) *Melville's Quarrel with God.* Princeton: Princeton University Press.

Thomsen, Mads Rosendahl (2008) *Mapping World Literature: International Canonization and Transnational Literatures.* London: Continuum Books.

Tocqueville, Alexis de (1990) *Democracy in America.* 2 vols., ed. Phillips Bradley, trans. Henry Reeve. New York: Vintage, 1990.

Turchi, Peter (2004) *Maps of the Imagination: The Writer as Cartographer.* San Antonio: Trinity University Press.

Vincent, Howard P. (1949) *The Trying-out of Moby-Dick.* Boston: Houghton Mifflin.

Wallerstein, Immanuel (1974) *The Modern World-System.* 3 vols. New York: Academic Press.

Weaver, Raymond (1921) *Herman Melville, Mariner and Mystic.* New York: George H. Doran Co.

Wegner, Phillip E. (2002) *Imaginary Communities: Utopia, the Nation, and the Spatial Histories of Modernity.* Berkeley: University of California Press.

Wellek, René (1963) *Concepts in Criticism.* New Haven: Yale University Press.

Westphal, Bertrand (2007) *La Géocritique: Réel, fiction, espace.* Paris: Minuit.

Williams, Raymond (1973) *The Country and the City.* Oxford: Oxford University Press.

— (1976) *Keywords.* Oxford: Oxford University Press.

— (1977) *Marxism and Literature.* Oxford: Oxford University Press.

Wilson, James C. (1981) "Bartleby": The Walls of Wall Street. *Arizona Quarterly* 37 (Winter): 335–346.

Wölfflin, Heinrich (1964) *Renaissance and Baroque,* trans. Katherine Simon. Ithaca: Cornell University Press.

Index

Lightning Source UK Ltd.
Milton Keynes UK
22 December 2010

164721UK00001B/47/P

9 780826 471512